NEW BRAIN, NEW WORLD

NEW BRAIN, NEW WORLD

How the Evolution of a New Human Brain Can Transform Consciousness and Create a New World

ERIK HOFFMANN

Australia • Canada • Hong Kong • India
South Africa •United Kingdom •United States

First published and distributed in the United Kingdom by:
Hay House UK Ltd, 292B Kensal Rd, London W10 5BE
Tel: (44) 20 8962 1230; Fax: (44) 20 8962 1239
www.hayhouse.co.uk

Published and distributed in the United States of America by:
Hay House, Inc., PO Box 5100, Carlsbad, CA 92018-5100
Tel: (1) 760 431 7695 or (800) 654 5126;
Fax: (1) 760 431 6948 or (800) 650 5115
www.hayhouse.com

Published and distributed in Australia by:
Hay House Australia Ltd, 18/36 Ralph St, Alexandria NSW 2015
Tel: (61) 2 9669 4299; Fax: (61) 2 9669 4144
www.hayhouse.com.au

Published and distributed in the Republic of South Africa by:
Hay House SA (Pty), Ltd, PO Box 990, Witkoppen 2068
Tel/Fax: (27) 11 467 8904
www.hayhouse.co.za

Published and distributed in India by:
Hay House Publishers India, Muskaan Complex, Plot No.3, B-2,
Vasant Kunj, New Delhi – 110 070
Tel: (91) 11 4176 1620; Fax: (91) 11 4176 1630
www.hayhouse.co.in

Distributed in Canada by:
Raincoast, 9050 Shaughnessy St, Vancouver, BC V6P 6E5
Tel: (1) 604 323 7100; Fax: (1) 604 323 2600

A catalogue record for this book is available from the British Library.

ISBN: 978-1-84850-827-9

Printed and bound in Great Britain by CPI Antony Rowe,
Chippenham SN14 6LH

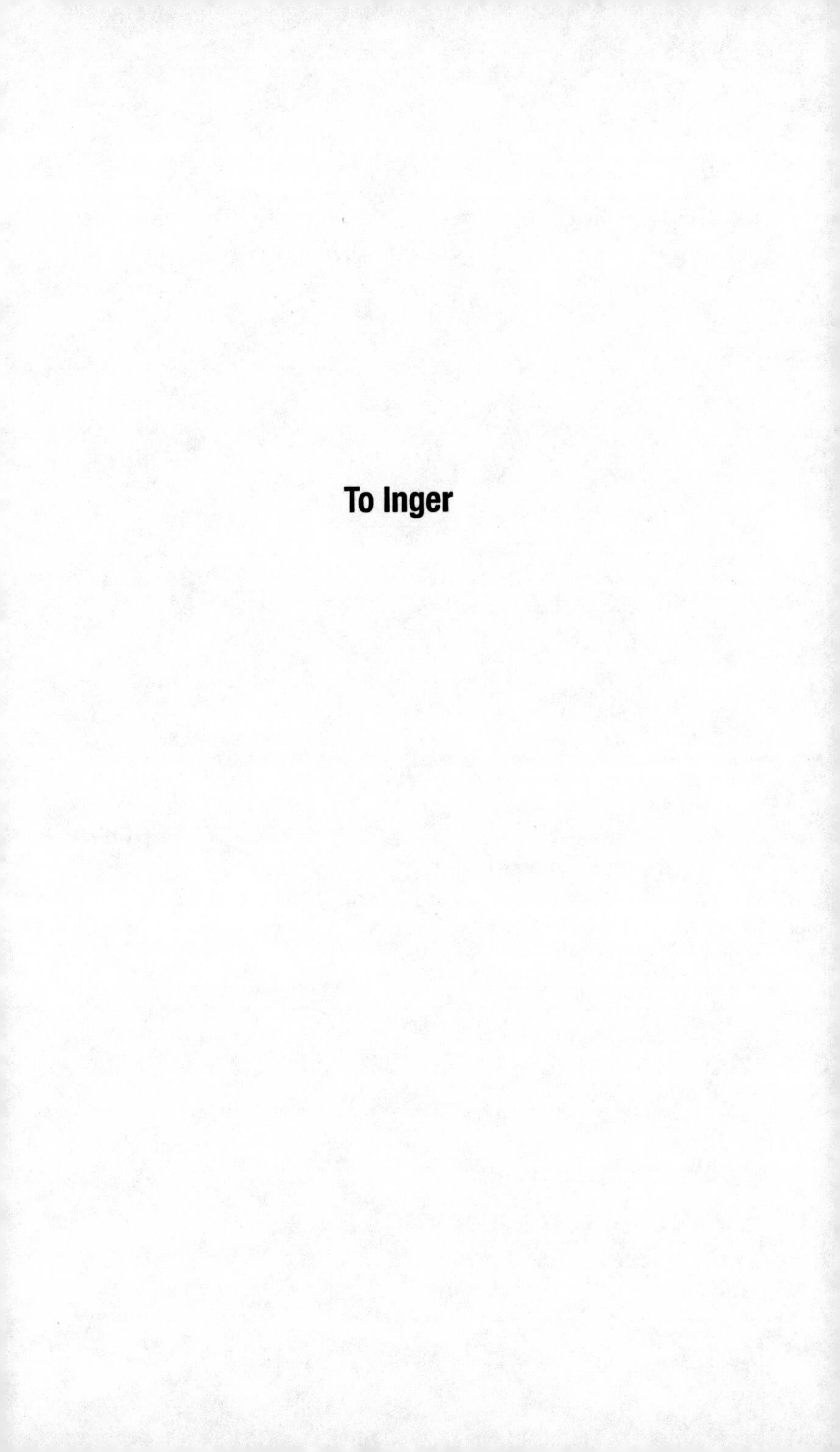

To Inger

CONTENTS

ACKNOWLEDGEMENTS

I am sure it wasn't a coincidence that I started writing this book in Goa, India, in January 2007. India and the people of India have given me so much inspiration. I have travelled to many ashrams, talked to many interesting people there and read many books by Indian masters. Thank you so much.

This book is dedicated to my partner, Inger Spindler, who has been invaluable to me in writing this book. She has been my primary discussion partner and supporter from day one and throughout the creation of this book. I have had numerous discussions with her on various subjects from the book. Her input on *kundalini* has been especially valuable, since she experienced a *kundalini* awakening at first hand in India in 2006. I have included an extensive interview with her on *kundalini* awakening and *samadhi*.

An early version of the manuscript was read, commented on and supported by a Danish colleague, Klaus Gormsen, and by an English friend, Steve Moses. Thank you.

I also want to thank Yatra da Silveira Barbosa, with whom I did an EEG study of ayahuasca in Brazil. I have included an interview with Yatra on ayahuasca and *kundalini* in the book. Also thanks to the Italian Nandan for her touching story about her loss and grief and her awakening in India.

I appreciate very much that Hay House UK selected my manuscript from many and had the courage and vision to publish it. I especially want to thank the editorial team, Amy Kiberd, Dan Benton and Lizzie Hutchins, for their advice and professional handling of the manuscript.

Also thanks to Danish engineer Per Gaarde-Nissen, who supplied and maintained the portable EEG brain-mapping equipment I used in all the studies described in this book.

PREFACE

When I was 13 years old I tried to stand on my head following instructions from a book I had just read about yoga and meditation. That book was my first encounter with spirituality. I was fascinated by its descriptions of yogis meditating in caves in the Himalayas and experiencing altered states of consciousness. Later, in my late teens, I picked up a book entitled *The Book of Life*, by a Danish philosopher and mystic called Martinus. This made a strong impression on me and further stimulated my interest in philosophy and psychology. I became insatiably curious to understand how brain functions were related to consciousness.

I started studying psychology at Copenhagen University in Denmark in 1965. After graduating as a master of psychology, I worked at Copenhagen University for some years as an assistant professor of psychology. I did research in psychophysiology and biofeedback, and in 1973 submitted a thesis entitled 'Biofeedback'. Biofeedback is a method that provides you with a signal (usually auditory or visual) informing you of the variations of a specific physiological variable, for example your brain waves. Through a simple process of trial and error, most people are able to learn to control, to a certain degree, a variety of physiological variables, including their brain waves, and this appeared to me to be a fascinating method of studying the relationship between brain activity and consciousness.

In the early eighties I spent some time at Rutger's University in New Jersey as a visiting assistant professor studying psychiatric patients using quantitative EEG (QEEG) methods, i.e. evaluating brain function by recording the electrical activity in the brain. I worked with the late professor Leonide Goldstein, a world-renowned authority on EEG, who taught me a lot and also became a personal friend. Later, in the nineties, I was research director for two years at Dr Arthur Janov's Primal Center in Los Angeles, where I studied the effects of feeling release (primal) therapy on patients' brainwave activity.

I remained fascinated by yogic masters and altered states of consciousness, and over the years I became involved in many scientific studies in this area. In Denmark and Sweden I studied the effects of Kriya Yoga meditation, and in Brazil I did a study on the effects of ayahuasca, a mind-expanding substance taken in the form of a herbal tea. In India, I studied brain changes in a group of people following a 21-day course involving, among other things, the 'transmission' of energy (called *deeksha* or *shaktipat*), and also had the opportunity to study a person very close to me before and after she had a *kundalini* awakening, which is a rush of energy surging from the base of the spine all the way up to the brain.

Most recently, in 2001, with my partner, the neurotherapist Inger Spindler, I founded the Mental Fitness and Research Centre (MFRC) in Copenhagen. At MFRC we have worked intensively with children with attention and behaviour disorders and adults with stress problems. In addition, we have conducted what we call 'high performance training' aimed at improving focus, presence and creativity.

This book is, in part, an attempt to summarize my lifelong study of human consciousness. Along the way, I have been inspired by many people. I have visited a number of ashrams in India, including Osho's International Commune in Poona, and the many discourses and writings of Osho have been of great value to me. Contemporary Western philosophers such as Eckhart Tolle and British physicist Peter Russell have also had a great influence on my thinking. If we look to the timeless wisdom conveyed by Eastern masters and some Western writers, I believe we already have all the knowledge we require about the mysteries of consciousness. What we need now is an understanding of how consciousness relates to the brain, and for that we require a lot more research. It is my hope that science will soon be inspired on a larger scale by the wisdom traditions of the world and will use the advanced technology now available to us in the study of human consciousness, thus bridging science and spirit.

INTRODUCTION

The physical sciences have come a long way in the past decades and technological progress has been astounding. New technologies have provided much-improved material standards and financial security in many parts of the world. However, calamities such as global warming, environmental pollution, terrorism, armed conflict and the exploitation and suppression of millions of people constitute such severe problems that if they are not corrected soon, our civilization and our planet are headed toward catastrophe. I do not doubt that the culprit behind these calamities is the human ego, with its greed, selfishness and lack of love and compassion for its fellow human beings. What is needed to create a better world is a change of consciousness.

Science, therefore, needs to direct its attention toward the more neglected inner world and put more effort into the study of human consciousness. In an essay called 'Deep Mind', Peter Russell writes:

> *Science has explored the world of space, time and matter and found neither evidence nor need for God. This would seem to have done away with religion. But the one realm Western science has not explored is the inner world of conscious experience. Those*

> *who have are the mystics, monks and yogis who have observed mind first-hand.*
>
> *(Russell 2003)*

There is no doubt that consciousness researchers can learn a lot from philosophical and spiritual traditions. The worldview of mainstream science is based on the following three premises: 1) humanity has reached the climax of its evolution; 2) humans are totally separate from each other, from nature and from the cosmos; 3) only the physical world exists. However, in this book I will argue that the human brain is in a state of organic evolution, which can result in global consciousness transformation, and that the physical world is only part of a total reality uniting humanity and all living creatures with the cosmos.

The first chapter briefly describes contemporary consciousness research and outlines the need for a shift away from the commonly held view that consciousness is a product of brain activity.

Chapter 2 looks at the brain's evolution. The human brain has undergone huge changes over millions of years. Currently, we have a double brain – a rational, verbal, analytical left hemisphere and an intuitive, non-verbal, creative right hemisphere. Most people seem to identify with the left hemisphere and repress, to a greater or lesser extent, the right one. Will our future evolution integrate these two cerebral hemispheres and what might be the implications for consciousness?

In Chapter 3, 'Expanding Human Consciousness', I describe my own research into altered states of consciousness. As mentioned earlier, I have studied the effects on the brain of meditation, the 'transfer' of energy, feeling release therapy and the drinking of ayahuasca.

An important driving force of consciousness development is the energy known as *kundalini*. In Chapter 4 we take a look at this and consider a case of *kundalini* awakening.

Awareness is another important driving force of consciousness development. This can be enhanced through meditation or through brainwave training (neurofeedback). I have worked for many years with neurofeedback and have found it quite effective for learning meditation and developing consciousness. This work is described in Chapter 5, where I also reflect on the efficacy of neurofeedback training for improving creativity, flow and intuition.

Chapter 6, 'A New Brain and a New Consciousness', looks at the possible future evolution of the brain. I believe that the human brain has not yet reached the peak of its evolution and is moving toward a higher level of functioning which I have termed the *New Brain*. This New Brain will support the experience of a higher level of reality, which in turn, as outlined in the final chapter, 'Toward a New World', will transform our world.

I admit that some of the theories and models put forth in this book are very bold and speculative, though I have tried to link as many of them as possible to empirical evidence, but I think it is important to speculate and put forth new hypotheses when the old ones become inadequate and obsolete. This is necessary for science to progress.

Unfortunately, many scientists have been reluctant to study consciousness, since it seems to be such a difficult and evasive subject. The tendency among research psychologists has been to concentrate on something more tangible and 'objective'. This has often been necessary in order to have their research funded, but it has led to most scientists getting to know more and more about less and less – not a satisfying position for people who want to know more about the universe, consciousness and the mystery of life.

It is my hope that this book will inspire consciousness researchers to adopt some of my hypotheses and make them the subject of rigorous scientific study. Remember Schopenhauer's words: 'All truths pass through three stages: they are ridiculed, they are opposed and finally they are regarded as self-evident.' By this, I do not mean to say that all of my hypotheses in this book will turn out to be true, but I am sure that some of them will. My only wish is to help science progress in the realm of consciousness and the human mind.

Please note: later chapters refer to six brainmaps, which do not display well in the black and white text of a book. These brainmaps are available at www.newbrainnewworld.com.

Chapter 1

THE SCIENCE OF CONSCIOUSNESS

Every two years since 1994, the University of Arizona in Tucson, USA, has held an international multidisciplinary conference entitled 'Toward a Science of Consciousness'. I attended the conference in 1996 and noticed that some of the attendees were asking very relevant questions such as 'How can consciousness emerge in a physical universe?' and 'Is consciousness an emergent property of brain processes or is it an inherent quality of the universe?'

Quite simply, scientists today cannot explain how consciousness works. How can science study an immaterial phenomenon such as consciousness? That is the problem!

The most popular current scientific methods for studying consciousness are neuroscience, cognitive science, philosophy and physics. Neuroscience is an old science, which tries to relate specific areas of the brain to specific psychological and cognitive functions, for example the left frontal-temporal areas to language functions. Some philosophers call that the 'easy problem' of consciousness and it is generally agreed that it has more or less been solved. A great number of brain functions, such as perception, language, learning and memory, can all now be explained in terms of natural brain science.

Cognitive science, on the other hand, is a relatively new and very popular discipline closely associated with computer science. Cognitive scientists work with computer models, robots, artificial intelligence, voice recognition and neural networks, to mention just a few areas. One of their goals seems to be to construct a robot or computer with a conscious mind. This is a hot issue among some cognitive scientists, who firmly believe that one day they will be able to construct a robot that simulates a human being so closely that you may be fooled into believing it has a conscious mind. (This is the 'Turing Test', named after the British mathematician Alan Turing.) Such a robot would, however, be a zombie, a machine that would act as if it were conscious but which would not have any subjective experience. It could pretend to have feelings, for example, but it would not experience them.

This is the so-called 'hard problem' of consciousness that has been outlined by Australian philosopher David Chalmers. Chalmers wonders how a physical system (the brain) can have an experience. Who is having the experience? Is the brain as a whole or part of the brain experiencing what is going on in the brain? And if so, how can a physical system like the brain experience a physical input? Why is there an experience at all? 'Why do these (brain) processes not go on "in the dark" without any subjective quality?', Chalmers asks (Chalmers 1996). This is the phenomenon that makes consciousness a real mystery.

As even Harvard psychologist Steven Pinker, who believes that thinking is just a form of computation and the brain an information processing system, admits that the hard problem remains a mystery.

Most scientists today, such as Pinker, support the theory of *reductive materialism*, which holds that consciousness is a by-product of brain activity and can be reduced to electro-chemical processes in the brain.

Nobel laureate Francis Crick (1995), believed by some to be the father of the scientific study of consciousness, says in his book *The Astonishing Hypothesis* that your joys, your sorrows and your sense of personal identity are simply the behaviour of a large assembly of nerve cells and molecules – a rather gloomy view.

Reductive materialists believe that if there is a sufficiently high level of brain organization and complexity, consciousness will naturally emerge from it. This 'emergence theory' has many supporters among contemporary scientists, even though they cannot explain *how* consciousness arises from brain activity. Some scientists argue that at our present stage of brain evolution we are not smart enough to figure it out, but eventually (when we have developed smarter brains) we will be able to explain the connection.

However, US brain researcher Andrew Newberg sees it otherwise, saying, 'The brain is not responsible for consciousness *per se*. It is just the vehicle by which consciousness can be manifested.' (Newberg 2004)

The Dalai Lama (2005) even suggests that the reductive materialists do not represent a scientific but rather a metaphysical position. He says that the view that all aspects of reality can be reduced to matter is as much a metaphysical position as the view that an organizing intelligence created and controls reality.

Even though most scientists do adhere to reductive materialism and the emergence theory of consciousness, more and more researchers are opening up to the possibility that consciousness is a natural aspect of the universe and life and precedes everything else.

One must bear in mind here that our conscious experience of reality is not a true reflection of the physical reality 'out there'. We do not and cannot know what exists out there. We only know that our senses are being stimulated by something in the external world that may give rise to a conscious experience. For example, when we look at a rose, an image of that rose is formed on the retinas of our eyes, giving rise to electrochemical processes in that area. From the retinas, electrical impulses are conveyed to the visual areas in the brain, and eventually electrochemical processes in the brain produce the experience of a rose. Those electrochemical processes are not identical to the rose; the reality we experience is only an interpretation by the brain of our external environment. It is not real, only virtual. That is why Indian masters call our experience of the outer world an illusion, *maya*, meaning that it is not what it seems to be.

DISTINGUISHING CONSCIOUSNESS FROM THE HUMAN MIND

Let me make my own position clear. I cannot imagine that consciousness emerges, as if by magic, at a certain point in brain evolution. In my view, consciousness has always been and always will be. It is a fundamental aspect of the universe that manifests in numerous forms. Probing the brain in order to find consciousness seems to me to be like taking a TV set apart in order to see where the pictures and sounds come from. Consciousness is not located in the brain; in fact, it has no location at all. As the German mathematician and philosopher Gottfried Leibniz said, 'If you could blow the brain up to the size of a mill and walk about inside, you would not find consciousness.'

Consciousness is omnipresent and eternal. What changes is the way it expresses itself in humans through the brain and its mind. The human mind is just a temporary construct and subject to continuous evolution. It is made up of both conscious and unconscious mental-emotional activity, which is dependent on the proper functioning of the brain. Therefore, the mind uses the brain in order to function.

Thus it is very important to discriminate between absolute, eternal Consciousness and the human conscious mind. In order to clear up any confusion, I will from now on use a capital 'C' when referring to eternal Consciousness and a lower-case 'c' when referring to the consciousness of the human mind. I see Consciousness as an eternal, immaterial and fundamental part of the universe and the mind as a form of Consciousness, a temporary creation that does not survive physical death. Consciousness does not emerge from matter as a result of sufficient brain mass and complexity, but the mind does. The mind develops as a form in the play of Consciousness with the cosmic energies.

As I mentioned earlier, ever since my late teens my personal view of the world has been greatly influenced by the Danish philosopher Martinus. Having undergone a deep spiritual experience, he claimed that the whole universe was a living manifestation of God. Perhaps his most important insight was that all living beings consisted of three elements, which he called X1, X2 and X3. According to this system, X1 is the core of the individual, the eternal Consciousness; X2 the ability to create through combining cosmic energies; and X3 the created form which is made up of cosmic energies, physical as well as non-physical. (Martinus 1939)

All we can say about X1, the eternal Consciousness, is that it has the ability to experience. According to Martinus, it can never be an object of further study. All we can study are the forms of Consciousness, how it manifests in matter and how it is transformed in living beings through their brains and nervous systems.

Eckhart Tolle has this to say about the eternal Consciousness:

> *So who is the experiencer? You are. And who are you? Consciousness. And what is consciousness? This question cannot be answered. The moment you answer it, you have falsified it, made it into another object... The subject, the I, the knower, without which nothing could be known, perceived, thought, or felt, must remain forever unknowable.*
>
> *(Tolle 2005: 242)*

Peter Russell writes in his eminent book *From Science to God* that modern science's inability to account for Consciousness will eventually push Western science into a paradigm shift:

> *I now believe that rather than trying to explain consciousness in terms of the material world, we should be developing a new worldview in which consciousness is a fundamental component of reality.*
>
> *(Russell 2002: 29)*

He goes on to say: 'God is universal. So is the faculty of consciousness. It is a primary quality of the cosmos, an intimate aspect of all existence.' (Ibid: 90)

Russell believes in *panpsychism*, meaning that Consciousness is in everything: 'The faculty of Consciousness must be present all the way down the evolutionary tree.' To illustrate this, he refers to a Sufi teaching: 'God sleeps in the rock, dreams in the plant, stirs in the animal, and awakens in man.' (Ibid: 114)

Instead of seeing Consciousness as a product of the brain, Russell suggests that the brain and nervous system are amplifiers of Consciousness, increasing the richness and quality of experience. He compares the brain to a film projector. The light in the projector represents Consciousness, the film is the mind and the pictures on the screen represent the experience. The lens in the projector corresponds to the brain and nervous system. Without the lens there will still be light on the screen, but the images will be much less sharp. That may be the way animals perceive the world, since they have a much less developed brain than humans.

Deep down inside, all living beings are connected to the same Consciousness that is God. Russell writes:

> ...*your* sense of I-ness is indistinguishable from *mine*. The light of consciousness shining in you, which you label as 'I', is the same light as I label as 'I'. In this we are identical. I am the light and so are you.
>
> *(Ibid: 82)*

THE MIND AND THE EGO

The mind is made up of all the mental-emotional activity going on in the brain, consciously as well as unconsciously. Emotion-driven compulsive thinking consolidates the mind.

In my terminology, the ego is the part of the mind that rules the person and with which the person is usually identified. It is sort of a false self created by unconscious identification with the mind.

The ego derives its sense of self through the person's past history and through the mental-emotional activities of their mind. It also identifies with external things such as education, work position, socio-economic status, relationships and possessions. These things, of course, have nothing to do with who a person really is. Eckhart Tolle says, 'Ego is the unobserved mind that runs your life when you are not present as the witnessing Consciousness, the watcher.' (Tolle 1999: 150)

The basic patterns of the ego are *resistance*, *control*, *power*, *greed*, *defence* and *attack*. It thrives on resistance. As Indian mystic Osho says:

> *...without fight the ego cannot exist for a single moment... But your cooperation is needed to keep the ego alive, and the cooperation is through fight, resistance ... [yet] you are such a tiny part of the cosmos that it is absolutely absurd to fight with it. With whom are you fighting? All fight is basically against existence.*
>
> *(Osho 2005: 33–4)*

There are indications that the ego is primarily associated with the left, thinking hemisphere. Neuro-anatomist Jill Taylor, who had a stroke in her left temporal lobe, offers supporting evidence for this view:

> *During the process of recovery, I found that the portion of my character that was stubborn, arrogant, sarcastic, and/or jealous resided within the ego center of that wounded left brain. This portion of my ego mind held the capacity for me to be a sore loser, hold a grudge, tell lies, and even seek revenge. Reawakening these personality traits was very disturbing to the newly found innocence of my right mind.*
>
> *(Taylor 2008: 145)*

In conclusion, the mind can be a very useful instrument and we cannot function without it. However, if we use it wrongly it can take on a life of its own and become very destructive.

It is also important to realize that the mind is only a tiny instrument of Consciousness at large…

BEYOND THE MIND: THE POWER OF CONSCIOUS THINKING

Since most people are identified with their mind, they are also identified with their thinking. Many believe that thinking represents the highest level of human consciousness. This is not the case. In fact, to my mind, thinking is highly overrated.

Beyond the thinking mind is a higher level of consciousness representing intuition, love, compassion and creativity. When this higher level is in charge of thinking, I call it *conscious thinking*.

Most thinking is not conscious – it is unobserved and uncontrolled and can become very destructive. The point about thoughts is that they are not immaterial phenomena without substance, but electromagnetic realities that influence mind, brain and body. They are 'food' for the mind: we become what we think. Therefore *what* we think is crucial.

Most people in Western societies are stressed and harassed by their racing minds, which are constantly thinking and chasing them around in the daytime and keeping them awake at night. It is therefore important that we learn to stop this busy mind and discipline our thinking.

We once had a client at our neurofeedback clinic, a lawyer who was troubled by stress. I said to her, 'In school they teach you how to think. In our clinic, however, we teach you how to stop thinking.'

She was baffled and replied, 'But without thoughts I would not exist.'

The idea that we would take away her thoughts was so frightening that she decided to give up the training.

Thoughts are valuable tools for the analytic mind to use to solve practical problems. However, as Eckhart Tolle points out in his bestselling book *The Power of Now*, most of us are preoccupied with an almost incessant inner dialogue. We are constantly talking to ourselves, continually assessing the events around us and the behaviour of ourselves and of others. Most of the time this dialogue is not conscious, although it is the foundation on which we perceive reality. It also affects our feelings, which are our body's reactions to our thoughts. Negative thoughts therefore create negative feelings, which drain us of energy and may lead to stress and disease.

It is an awful burden not to be able to stop thinking, but most of us do not even realize this, since it is a very common condition. The problem is that up to 90 per cent of our thoughts are automatic – they go round and round like old records, more or less unconsciously. For the most part we are not using them; they are using us. And when we identify with them, we are letting them take control of us.

How can we regain control? It is very difficult to stop compulsive thinking. Saying you want to stop doesn't work. You cannot stop a thought with another thought; the order must come from a higher level. So you must raise your level of consciousness one step up and imagine that you are observing both your thoughts and the thinking process itself from a higher level. Try to be 'the silent watcher on the hill', as Osho puts it.

There are different methods by which you can raise your consciousness 'above' the thinking level. Some, such as meditation and brainwave training, are described in this book. In neurological terms, watching from a higher level probably means that the frontal lobes are observing and controlling everything that goes on in other parts of the brain. More on this later.

The point is that observing a thought not only makes you aware of the thought but also makes you conscious of yourself as a witness to the thought. A new dimension of consciousness has stepped in. The habitual, compulsive thoughts have lost their power and will gradually subside. Then, as you learn to stay conscious in the present moment, you can decide when to think.

In that state you are much more awake, aware and present than you are when you are identifying with your thoughts. This is eloquently expressed by Eckhart Tolle:

The moment you start watching the thinker, *a higher level of consciousness becomes activated. You then begin to realize that there is a vast realm of intelligence beyond thought, that thought is only a tiny aspect of that intelligence. You also realize that all the things that truly matter – beauty, love, creativity, joy, inner peace – arise from beyond the mind. You begin to awaken.*

(Tolle 1999: 14)

Chapter 2

THE EVOLUTION OF THE HUMAN BRAIN

What will it take for us all to awaken? I feel certain that for humanity as a whole to attain a higher consciousness, the brain must evolve to a higher level. The basis of a higher state of human consciousness (not to be confused with eternal Consciousness) must be biological.

I also feel that if we can construct brain models which can explain how consciousness, unconsciousness and higher consciousness relate to the brain, it will give us a much deeper understanding of these concepts.

So let's first look at the evolution of the human brain.

THE TRIUNE BRAIN: THREE LEVELS OF EVOLUTION

US brain researcher Paul MacLean has developed a captivating model of brain structure and evolution which he calls the *Triune Brain* (*see Figure 2.1*).

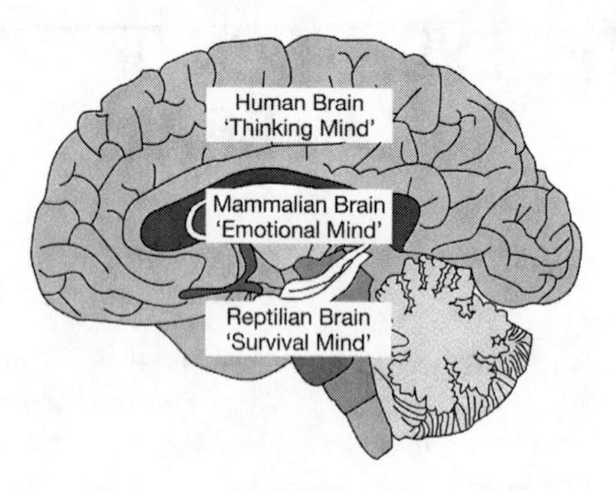

Figure 2.1: The Triune Brain

We actually have three brains, one on top of the other, each corresponding to a separate major evolutionary step. The oldest brain, the reptilian brain, which we share with fish and reptiles, is comprised of the spinal cord, the brainstem and the midbrain, including the hypothalamus. This is the location of the neural mechanisms involved in basic needs associated with reproduction and self-preservation: heart regulation, blood circulation, respiration, food intake, sexual and rage behaviour.

A much later structure in brain development is the limbic system, which we share with mammals. This is comprised of the amygdala, the hippocampus and the septum. It is involved in complex emotions and memory functions and is the primary seat of feelings.

Finally, surmounting the rest of the brain, there is the neo-cortex, which is clearly the most recent evolutionary brain structure. It has been estimated that it consists of some 100 billion neurons (brain cells). However, a Danish professor, Bente Pakkenberg, has calculated the number to be around 20 billion, with men having around 4 billion more than women. With the growth of this massive neo-cortex, a variety of cognitive functions developed, providing us with the ability to think, analyse and plan for the future.

According to psychologist Arthur Janov, these three interconnected brains – brainstem, limbic system and neo-cortex – are the seats of what he calls the 'survival mind', the 'feeling mind' and the 'thinking mind' respectively. Thus, we have three different levels of consciousness based on specific brain structures that evolved over eons of time.

MacLean's Triune Brain model (MacLean 1973) underscores that man within his skull has three separate brains, each with its own distinct capacities. Even if they are heavily interconnected they do not always co-operate. Normally the highest level, the neo-cortex, should be able to control the lower levels. However in case of an affective crime we often see how the limbic lower level can 'hijack' the upper level.

THE DOUBLE BRAIN: TWO BRAINS, TWO MINDS

As well as the three brains arranged vertically within our skull, we also have two brains arranged horizontally, namely the left and the right cerebral hemispheres. The left brain controls the right side of the body and the right brain controls the left side of the body.

Early studies on the effects of one-sided brain lesions determined that in the majority of people language functions were located in the left brain. It was not until much later that it was discovered that the right hemisphere also had specific cognitive and psychological functions (*see Table 2.1*).

Table 2.1: Hemispheric Specialization

Left Hemisphere	Right Hemisphere
Logical	Holistic
Analytical	Intuitive
Verbal	Integrative
Quantitative	Synthesizing
Organized	Interpersonal
Sequential	Empathic
Deliberate	Spontaneous

The two hemispheres of the brain are usually subjected to the same external stimuli but process them in different ways. The left hemisphere, which is the dominant hemisphere in right-handed and in most left-handed individuals, operates in a logical, analytic, computer-like fashion which makes it suitable for language processing and mathematical operations. The right hemisphere, in contrast, specializes in gestalt perception and the synthesis of incoming information, which makes it superior in dealing with audio- and visual-spatial tasks such as the perception of music, orientation in three-dimensional space and the recognition of faces. The right hemisphere also plays an important role in the processing of images and feelings. According to neurophysiologist Rhawn Joseph, 'The right brain imparts meaning, context, sincerity and intent to the verbal analytic processes of the left brain.' (Joseph 1992: 79)

The case of Jill Bolte Taylor, a Harvard-trained neuro-anatomist who had a stroke in her left temporal lobe at the age of 37, throws more light on how the two hemispheres work. The stroke eliminated most language functions and affected large parts of the left hemisphere. Afterward, Taylor was able to watch, from her well-functioning right hemisphere, how her left mind gradually deteriorated. She lost the ability to talk and could only understand a few words if they were spoken slowly. Eight years later she had totally recovered and she told her fascinating story in the book *My Stroke of Insight*. She describes her left hemisphere as follows:

> *My left mind is the tool I use to communicate with the external world. Just as my right mind thinks in collages of images, my left mind thinks in language and speaks to me constantly... Via my left brain*

> *language center's ability to say 'I am', I become an independent entity separate from the eternal flow. As such, I become a single, a solid, separate from the whole. Our left brain truly is one of the finest tools in the universe when it comes to organizing information. My left hemisphere personality takes pride in its ability to categorize, organize, describe, judge, and critically analyze absolutely everything.*
>
> *(Taylor 2008: 142)*

Taylor goes on to say about her right brain:

> *My right hemisphere is all about right here, right now. It bounces around with unbridled enthusiasm and does not have a care in the world. It smiles a lot and is extremely friendly. In contrast, my left hemisphere is preoccupied with details and runs my life on a tight schedule... To my right mind character, there is no judgement of good/bad or right/wrong, so everything exists on a continuum of relativity. It takes things as they are and acknowledges what is in the present.*
>
> *(Ibid: 139)*

The Split Brain: Disconnecting the Left and Right Brain

That the two cerebral hemispheres are functionally different and in many ways behave like two independent brains was convincingly demonstrated by the breakthrough work of Nobel laureate Roger Sperry and his colleagues with 'split-brain' patients at California Institute of Technology in Los Angeles in the 1960s (Sperry 1974). In split-brain patients the cerebral hemispheres have been surgically disconnected from each other. This operation used to be carried out to treat intractable epileptic seizures (with success). Following a split-brain operation, each hemisphere appears to function independently, with its own perceptions, cognitions and memories, and even its own values, will and mind.

This can give rise to some inter-hemisphere conflict which, since the left hemisphere controls the right hand and the right hemisphere the left hand, may be seen in the behaviour of a split-brain patient's hands. One such patient complained that sometimes when he was shopping in the supermarket, his right hand would take items from the shelves and put them in the shopping trolley, but then his left hand would put them back on the shelves and take other items instead.

Another curious story was about a patient who would light a cigarette with his right hand only to find his left hand taking the cigarette and putting it out. Obviously the patient's left hemisphere wanted to smoke, while his right hemisphere wanted to quit.

This phenomenon, called *alien hand syndrome*, where one hand – usually the left – takes on a mind of its own, is often seen in split-brain patients immediately after the operation.

Mostly, however, the behaviour of split-brain patients appears to be quite normal. Nevertheless, laboratory tests have clearly showed subtle handicaps in their perceptions and behaviour. If a split-brain patient is shown an image or an object in their left visual field (connected to the right hemisphere), they will not be able to name it verbally. This is because the left, speaking hemisphere has no access to the information because it is disconnected from the right, visual hemisphere. However, the same patient will be able to retrieve the perceived object with their left hand from a bag filled with many other objects. Obviously the right hemisphere has the information about the object while the left has not.

It is interesting to note that split-brain patients show strongly reduced creativity, as demonstrated by German-born UCLA researcher Klaus D. Hoppe, and that, as the US neurosurgeon Joseph Bogen notes, 'the patients' fantasy life is reduced, or even absent' (Bogen and Bogen 1969). In addition the content of their dreams becomes less symbolic and more everyday. These changes are probably due to the left hemisphere's lack of access to the right hemisphere.

The Corpus Callosum*: Integrating the Left and Right Brain*

The right and the left hemispheres of the brain are connected by the *corpus callosum*. This consists of more than 200 million nerve fibres, which makes it one of the largest structures in the brain. In addition there is a smaller nerve cable (with an estimated 50,000 fibres), called the *anterior commissure*, connecting the left and the right temporal lobes at the front of the brain.

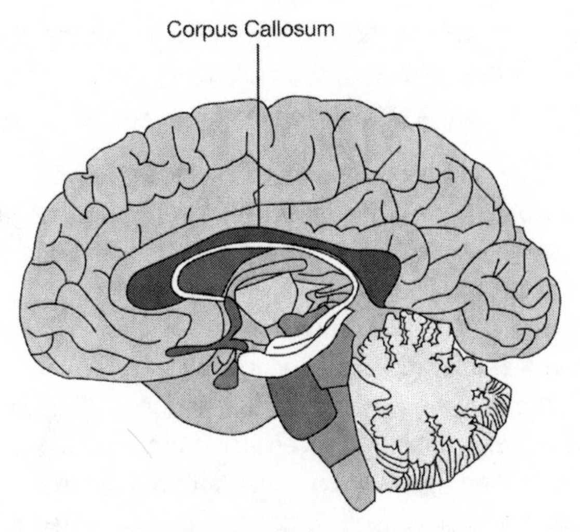

Figure 2.2: The Corpus Callosum

The most important and obvious function of the *corpus callosum* is to coordinate and integrate motor activity, sensory perception and thinking in the two hemispheres. If, for example, you ask a person to move their left hand, this verbal request is understood by their left hemisphere and its intent to move the left hand is then transferred, via the *corpus callosum*, to the right hemisphere, which executes the order.

Most of what we call thinking is done in terms of words and concepts and takes place in the left hemisphere, as this is dominant for language function. Thus, what I call *the thinking mind* resides primarily in the left hemisphere. However, the right hemisphere can also contribute to thinking. The right brain usually thinks in terms of pictures, images, sensations and feelings. During the thinking process the two hemispheres will interact via the *corpus callosum*, the left brain contributing with words, the right brain with images, sensations and feelings.

Many consciousness researchers believe that consciousness is totally dependent on language. I agree that language is important for thinking; however, it is not a prerequisite for consciousness. The way I see it, language functions in the left hemisphere work as a kind of 'operating system' for the human mind. Words and concepts are not the content of the mind; they are only signposts. Language, I suggest, is a system of 'file names' necessary in order to access the content of the mind, which consists mainly of sensations, images and feelings and is therefore mostly related to the right hemisphere and to the deeper brain structures such as the limbic system.

I believe that when we think, there is an almost constant dialogue going on between the left and the right hemispheres via the *corpus callosum*. This inner dialogue has been used as the main criteria for human consciousness (Jaynes 1976). The German Buddhist philosopher Herbert Guenther called consciousness 'awareness of awareness'. This fits in well with my theory that human consciousness as we know it emerges when the left hemisphere becomes aware of the right and the right hemisphere becomes aware of the left, and when the two hemispheres can respond to each other.

There are some puzzling results from split-brain operations regarding the function of the *corpus callosum*. Neurosurgeon Joseph Bogen characterizes split-brain patients' social behaviour following the operation like this:

> *One of the most remarkable results is that in ordinary social situations the patients are indistinguishable from normal in spite of the cutting of over 200 million nerve fibers. Special testing methods are needed to expose their deficits.*
>
> *(www.cco.caltech.edu/~jbogen/text/onesothe.htm)*

Many scientists have wondered, therefore, why such an impressive number of connections between the hemispheres is needed.

In fact some people are born without a *corpus callosum* and sometimes also without an anterior commissure. One famous example is US citizen Kim Peek, an autistic savant who has been intensively studied because of his phenomenal memory. He became the model for Dustin Hoffman's role when he played an autistic savant in the movie *Rain Man*.

Patients with a missing *corpus callosum* have problems evaluating their own state of emotional arousal. This is probably because the left, speaking hemisphere has no access to the right, feeling hemisphere. In a personality disorder called alexithymia, patients have similar difficulties in identifying and describing their feelings and I suspect that alexithymic persons also have problems with information transfer between the left and the right hemispheres.

In Alzheimer's disease, brain scans have shown significant reductions in the cross-sectional area of the *corpus callosum*. As the disease progresses, there is a decline in inter-hemispheric connectivity. It is interesting that this degeneration can be measured by electroencephalography, since the brain waves in the left and right hemispheres become less coordinated and coherent.

There is still a lot to uncover about the *corpus callosum*. There are some indications that the fibres in the rear part (connecting the sensory and association areas of the two hemispheres) are the most active and that severing them has the most obvious behavioural consequences. Bogen writes:

> *Apparently the posterior fourth of the corpus callosum, if left intact, can prevent the appearance of the entire disconnection syndrome as now conceived... For the remainder (about 150 million nerve fibers) no more is known now than 25 years ago.*
>
> *(Bogen and Bogen 1969: 4)*

The reason for this could be because this is the only part of the *corpus callosum* that is fully developed and active in most adults.

This hypothesis is in accordance with the fact that the neo-cortex, the part of the brain that has developed most recently, first matures in the rear (occipital-parietal) areas and much later in the frontal areas.

Numerous observations in my own research have shown that while the left and the right parietal lobes in the rear part of the brain easily affect each other (via the *corpus callosum*), this is not the case with the frontal lobes. This again leads me to believe that the neurons in the frontal part of the *corpus callosum* are not open for communication because they are not mature enough.

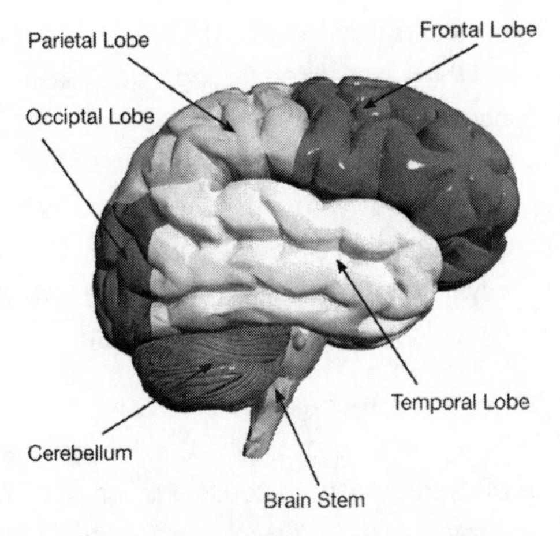

Figure 2.3: The Brain Lobes of the Neo-cortex

These findings and speculations leave me with the suspicion that this huge nerve bundle is meant to serve a higher purpose in the integration of the two hemispheres and in the development of a more evolved consciousness. I believe that at some point in future brain evolution the *corpus callosum* will be fully mature and there will be optimal connectivity between the two hemispheres. When this happens, it will provide the neurological infrastructure necessary for a full integration of the hemispheres and the emergence of a higher consciousness.

The Right Brain and the Unconscious

In his book *The Right Brain and the Unconscious,* Rhawn Joseph outlines his view that the left hemisphere is the seat of primary consciousness. Just below that there is a pre- or subconscious level associated with the right hemisphere.

Joseph believes that this is where long-forgotten childhood memories and their associated feelings are stored. Primary unconsciousness, however, resides deep in the brain's limbic system. So, if consciousness is seated in the left hemisphere, access to the deep unconscious must be facilitated by neural connections from here, either directly to the limbic system or (more likely) via the *corpus callosum* and the right hemisphere to the limbic system.

There has been a debate as to whether the right hemisphere is conscious or not. Joseph suggests that it is the seat of a secondary unconscious, or what he calls 'an ancient awareness'. However, there is ample evidence from split-brain research that the right brain is conscious. Roger Sperry writes:

> *Although some authorities have been reluctant to credit the disconnected minor hemisphere even with being conscious, it is our own interpretation, based on a large number and variety of non-verbal tests, that the minor hemisphere is indeed a conscious system in its own right, perceiving, thinking, remembering, reasoning, willing, and emoting, all at a characteristically human level, and that both the left and the right hemisphere may be conscious simultaneously in different, even in mutually conflicting, mental experiences that run along in parallel.*
>
> *(Sperry 1974: 11)*

During the first couple of years of life, before language evolves, the young child primarily operates through the right hemisphere. Thus, memories of possible abuse and traumatic events during early childhood are stored on the right side (Joseph 1992). When the child grows up and language functions develop (usually) in the left hemisphere, that hemisphere becomes the primary seat of consciousness. This change of dominance from the right to the left side of the brain seems to be firmly established during puberty (12–14 years). In this period the frontal lobes also take a significant maturational leap forward.

Following the shift in hemispheric dominance to the left side, the *corpus callosum* is still very immature and left-hemisphere consciousness will therefore have little or no access to the early memories of the right hemisphere. This is probably why early traumas are very hard to recover in psychotherapy – they remain secrets of the right hemisphere, though they may still exert an unconscious influence on the person's behaviour, possibly throughout life. We must also assume that what psychotherapists call the 'inner child' resides (and hides) in the right hemisphere.

It is important to emphasize that the right hemisphere is the original brain from which Consciousness operates. This can be seen in young children below two years of age. According to Harvard psychologist Julian Jaynes, it was also the case in ancient humans thousands of years ago. Thus, the left hemisphere appears to be a later evolutionary development. It even forms later *in utero*.

We know that the left hemisphere is superior in many respects and is necessary for handling most daily activities. However, as we shall see later, it sometimes seems that, at least in the Western world, it plays its intellectual games without much reference to the older evolutionary brain systems – it seems to rule as if it had no body. It is in fact more isolated from the deep brain structures than the right hemisphere, which is much more integrated with the limbic system, and hence the feelings, and with the rest of the body.

I cannot overemphasize the importance of the right hemisphere when it comes to life quality. When an individual looks at themselves and the world, they look through both the left and the right brain. While the left hemisphere supplies the facts and details about an experience, the right hemisphere contributes the background feeling that makes the experience real and significant. So, if a person has proper access to their right hemisphere, they are bound to have a much more profound experience of whatever they are experiencing – and are going to feel much more whole and alive.

In cases where people have little or no access to their right hemisphere, it is usually because it has been blocked or repressed, possibly because of trauma or severe stress early in life. Again, more about that later.

THE FRONTAL LOBES: THE FOURTH LEVEL OF BRAIN EVOLUTION

As we have just seen, the brain has evolved over millions of years through three separate stages of development: the reptilian, mammalian and human brains, supporting survival, emotion and thought respectively.

Most recently, the frontal lobes have developed. These take up almost one third of the human cortex (29 per cent according to neuro-anatomist Korbinian Brodman), but until about 50 years ago scientists didn't really know what their functions were and they were called the 'silent areas'. That was why such a radical operation as a lobotomy was allowed to be performed on certain patients (for example schizophrenia and anxiety patients), severing the frontal lobes from the rest of the brain. However, it has since been discovered that the frontal lobes, especially the area just behind the forehead (the prefrontal cortex), are the seat of some very important brain functions.

The frontal lobes are the sites of voluntary functions, the executors of intentions, motives and plans formulated with the aid of speech. The frontal cortex also directs attention and concentration. It is responsible for the control, interpretation and integration of subcortical brain activity, including feelings.

Neurophysiologist Elkhonon Goldberg has called the prefrontal cortex the CEO of the brain (Goldberg 2001: 23). It is the leader or conductor that oversees and controls all other parts of the brain to which it is neurologically connected.

Several Indian masters have claimed that the frontal cortex in most humans is in a sleeping state and needs more energy to awaken. There are also indications from recent research into meditation and higher states of consciousness that under the right circumstances the frontal cortex can function at a much higher energy level, generating very fast brainwave frequencies called gamma waves (*see Chapter 3*).

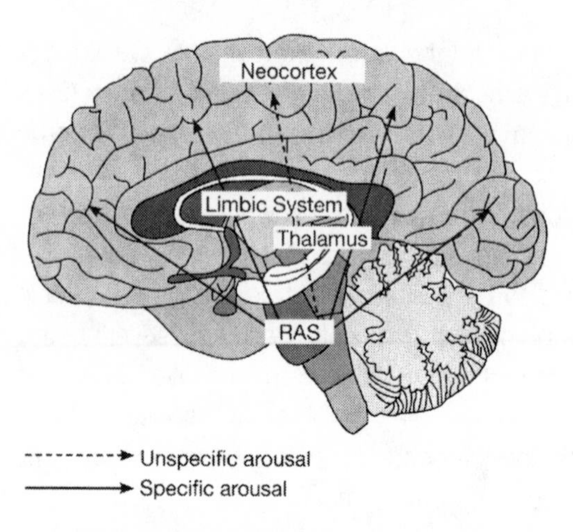

Figure 2.4: Arousal Mechanisms in the Brain

Some very important functions of the frontal lobes are the regulation of attention and wakefulness. It has been demonstrated that the prefrontal zones of the cortex have rich nervous connections with the rest of the brain, including the thalamus and the reticular activating system (RAS), which is located in the upper part of the brainstem and may be the physiological mechanism responsible for repression. When unpleasant or traumatic feelings activate the RAS and approach the frontal lobes for conscious integration, increased prefrontal activity may suppress these feelings. Some preliminary studies I have done at the Mental Fitness and Research Centre support the theory that increased activity in the left prefrontal area can suppress painful feelings.

There is another mechanism in the brain that can suppress feelings. Since feelings are primarily mediated by the right hemisphere, an overly active left hemisphere may prevent them from reaching the left hemisphere and hence the primary consciousness.

It has been suggested that the frontal lobes stand guard on either side of the *corpus callosum* to determine what may pass through. Neurophysiologist Rhawn Joseph says:

> *The frontal lobes are in essence, the senior executives of the brain, ego and personality. They control behavior, attention and information processing throughout the brain via inhibition, suppression, and censorship... They may act to prevent information from crossing over the 'psychic corridor' (the corpus callosum) via inhibition. Thus, it can be said that they engage in censorship and stand guard on either side of the* corpus callosum *to determine what may pass.*
>
> *(Joseph 1992: 66)*

Whether or not this hypothetical mechanism can be verified, it seems likely that the frontal lobes play an important role in regulating inhibition and communication via the *corpus callosum*. Therefore they seem to be responsible for creating a balance between the left thinking/intellectual mind and the right feeling/intuitive mind.

Chapter 3

EXPANDING HUMAN CONSCIOUSNESS

In order to understand human consciousness and its relationship to the brain, I have found it important to study how so-called altered states, or non-ordinary states, of consciousness affect the brain. By 'altered states of consciousness' I mean states that deviate from normal daily consciousness, such as sleep and dream states, hypnosis, meditative states and states induced by mind-expanding substances. My own research has primarily involved studying these states by means of quantitative electroencephalography (QEEG) and brain-mapping.

MEASURING THE BRAIN'S ELECTRICAL ACTIVITY (EEG)

The brain and its billions of cells operate on electricity. The brain is electrically active day and night throughout life.

When large groups of neurons in the brain are coordinated or coherent (working in synchrony), small electrical potentials build up to form so-called brain waves.

The electrical signals can be measured by electrodes placed on the scalp using conductive paste. After amplification, they are fed to a computer and displayed on the screen. This method of brainwave measurement is called electroencephalography (EEG).

The more neurons that are working in synchrony, the higher the voltage or amplitude of the electrical oscillations measured in microvolts. The faster the neurons work together, the higher the frequency of the oscillations measured as cycles per second or Hertz. These two parameters, *amplitude* and *frequency*, are the primary characteristics of brain waves.

If the EEG amplitudes are quantified at different frequencies, the process is called *quantitative electroencephalography* (QEEG), which is the method I used in my own studies.

Five Types of Brain Wave

Brain waves may be divided into five different categories according to their frequency:

- The slowest waves, called *delta waves*, are dominant during coma and deep sleep.
- The next slowest, *theta waves*, are associated with drives, emotions, trance states and dreaming.

- In the middle of the frequency band are the *alpha waves*. They are the prime indicators of conscious attention and represent a sort of 'gate' between the outer and inner world and between the conscious and the unconscious.
- *Beta waves* are fast waves indicating an aroused, mentally active and concentrated state.
- Finally, we have the fastest waves, called *gamma waves*. They reflect intense focus and alertness. As we shall see later, they are also found in some higher states of consciousness.

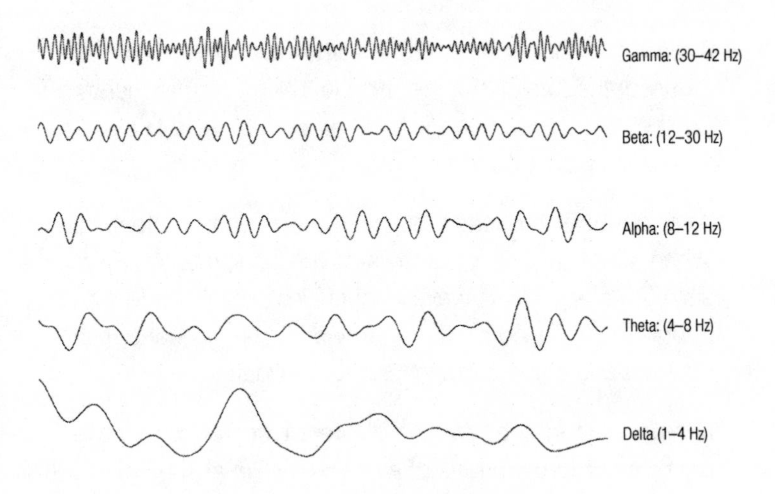

Figure 3.1: Five Types of Brain Waves

Both delta and theta waves reflect unconscious states, whereas alpha, beta and gamma waves indicate conscious states.

Table 3.1: Brain Waves, Frequencies and Functions

Unconscious		Conscious		
Delta	**Theta**	**Alpha**	**Beta**	**Gamma**
0.5–4 Hz	4–8 Hz	8–13 Hz	13–30 Hz	30–42 Hz
Instinct	**Emotion**	**Consciousness**	**Thought**	**Will**
Survival Deep sleep Coma	Drives Feelings Trance Dreams	Awareness of the body Integration of feelings	Perception Concentration Mental activity	Extreme focus Energy Ecstasy

A basic principle of the EEG is that the more active or aroused the brain, the higher the brainwave frequency. On the other hand, during states of drowsiness and sleep, slow brain waves prevail.

Also, the EEG frequency reflects the direction of attention. When attention is directed towards the inner, mental world, slow theta and alpha wave frequencies prevail, and when attention is focused on the outer world, fast frequencies in the beta and gamma range tend to dominate.

It is possible to see how the different types of brain wave are distributed by means of a brain map. In all the studies referred to in this chapter, I used a portable eight-channel EEG brain-mapping apparatus. Coloured brain maps were constructed in order to give a quick overview of the functional state of the brain.

A trained electro-encephalographer can tell from a quick look at a brain map whether, for example, there is too much theta activity in the frontal lobes, too little alpha activity in the rear part of the brain and a balance of activity between the left and the right sides of the brain. (*See www.newbrainnewworld.com.*)

Brainwave Frequencies and Levels of Consciousness

One purpose of my research was to determine how the levels of consciousness were characterized by different brainwave frequencies.

Brainwave frequency reflects levels of consciousness associated with different brain structures and brain functions. While the older, deep structures of the brain are associated with slow brain waves (delta and theta), the more recent structures on top of the brain (neo-cortex) are associated with faster brainwave frequencies (alpha, beta and gamma). As a general rule, the brainwave frequency reflects the maturity of the brain structure. The faster the frequency, the more mature (and the more active) the brain.

Table 3.2: Brain Structure and Levels of Consciousness Reflected by Brainwave Frequency

Brain in Evolution	Brain Structure	Level of Consciousness	Brainwave Frequency
Reptilian	Brainstem	Survival	Very slow (Delta)
Mammalian	Limbic system	Emotion	Slow (Theta)
Human	Neo-cortex	Thought	Medium to fast (Alpha / Beta)
Superhuman	Frontal lobes	Focus / Presence	Medium to very fast (Alpha / Gamma)

In a newborn baby, only the reptilian part of the brain is fully developed, providing the baby with the basic physiological functions necessary for survival. At this age, slow delta waves are the most characteristic rhythms of the brain. These reflect the survival and reproductive functions of the deep brain structures and are the signatures of instinct and survival.

The limbic system matures in the young child, and up to the age of six, slow theta waves predominate. They are indicative of limbic system functioning and their appearance in the EEG is closely associated with emotional activity. Thus theta waves reflect activity at the brain's emotional (limbic) level and are the signature of feelings.

As the neo-cortex matures in the child, the theta rhythms are gradually replaced by the faster alpha and beta waves, reflecting the development of the neo-cortex and the 'thinking mind'. By the age of six or seven, most children have well-established alpha rhythms, at least in the rear part of the brain. Since the frontal lobes mature later than the rest of the neo-cortex, theta waves often prevail in the frontal regions until the age of 12 to 14, at which time the slightly faster alpha waves become dominant. This frequency increase in the frontal lobes correlates with the child's ability to cope with and control emotions.

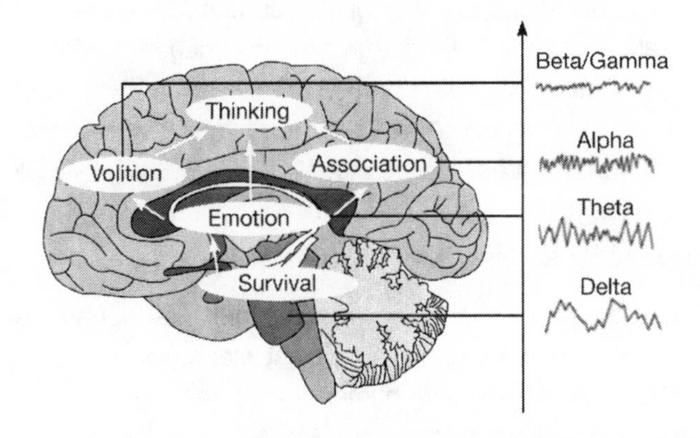

Figure 3.2: Brainwave Frequencies Reflect Brain Functions and Levels of Consciousness

DEEP LEVELS OF CONSCIOUSNESS: SLEEP AND DREAMS

There are three basic levels of consciousness, each associated with its own specific brain structure: deep sleep, dream sleep and the waking state. Sleep and dream sleep are unconscious states, and are the most basic altered states of consciousness we know.

It is interesting to see what happens in the brain when we go to sleep at night. If we have been watching TV, it will probably have stimulated our brain to generate fast beta waves. When going to bed afterwards, we lie down, close our eyes and relax. In that state of awake mental relaxation, our brain will normally produce alpha waves, the signature of the brain's idle state. Next, we become drowsy and lose contact with the environment. Now the alpha waves drop down and theta waves tend to appear. In this state (the hypnagogic state), unconscious or subconscious material is likely to surface in the form of mental imagery, fantasies and dreamlike pictures. In my own research I have found that the more emotional conflicts a person has, the more theta waves appear in this state. Then eventually we fall into deep sleep, where the brain produces slow delta waves. This is a state of deep unconsciousness.

During the process of falling asleep it looks as though the brain descends through the different stages of evolution. At first the neo-cortex (human brain) falls asleep and the alpha activity disappears. Next, the limbic system (mammalian brain) falls asleep and the brainwave activity becomes even slower. Finally, only the delta waves of the deep brain structures (the reptilian brain) are active, supporting the vital functions of the body. Thus during the night we regress down through the levels of consciousness and experience their associated realities. This seems to be a necessary restorative and healing process for the body and the mind.

We experience the same process in reverse when waking up. From deep sleep, the limbic system wakes up and we start dreaming. This is of course an unconscious state. Nevertheless it will cause an activation of the EEG (especially in the right hemisphere), which may look as if we are awake. During the dream state, theta waves often prevail, reflecting emotional activity in the limbic system of the mammalian brain. Eventually the whole neo-cortex will wake up and start producing alpha or beta waves.

When we go to bed at night, in a sense we lie down beside a crocodile and a horse. These animals represent our reptilian and mammalian ancestors in the brain. If we don't have a friendly relationship with these animals, our sleep may be disturbed by violent dreams and nightmares. In that case we shouldn't be surprised if we wake up the next morning exhausted and confused.

It seems very likely that if we ignore our emotional problems while awake, they will haunt us in our dreams. Such problems are usually associated with the limbic system, which is active during dreaming. In the dream state the neo-cortex seems to be more or less asleep and therefore cannot repress emotions, which may then appear in our dreams. The moral is that the more you deny your emotional problems, the more likely they are to emerge in your dreams and disturb your sleep.

Why do we visit the dream states and dreamless sleep states of our unconscious every night of our lives? In fact sleep and dreams make up almost one third of our lives. Obviously the body needs the rest and restitution it gets during sleep. Also, I believe, the mind needs to dream to release residues of daily stress and emotional conflict. This is supported by research which has shown that the more stressed people are, the more they dream (Foulkes 1966).

Psychoanalysts such as Freud and Jung have made many attempts to analyse dreams and make some sense out of them. I believe, however, that most dreams are just mental/emotional garbage. They are distorted messages from the unconscious and therefore, to a large degree, useless. It is the mind unwinding trying to get rid of unfinished business from the awake state, mostly of an emotional nature. Osho says:

> *The reality of a dream you will never find. It is simple garbage, it is really the unwinding of the mind… The dream is a great help, a cleaning … it keeps you sane … but dreaming has no other significance.*
>
> *(Osho 2001)*

However, since dreams reflect unconscious activity, it is sometimes possible to identify emotional problems through the interpretation of dreams. Osho acknowledges that dreams are messages from the unconscious. However, he advises not to analyse them but just observe them with full awareness. This can be done as an exercise in the morning or whenever you wake from a dream. If you are skilled enough, this awareness can carry over so you know that you are dreaming when you dream. This is called *lucid dreaming*.

There are indications that the more you succeed in calming your mind and emotions, through meditation or otherwise, the less you dream. Osho says:

> *When a person represses nothing, dreams disappear. So a Buddha never dreams. If your meditation goes deep you will immediately find that your dreams are becoming less and less and less.*
>
> *(Ibid: 12)*

If the function of dreaming is unwinding the mind, reducing stress and cleaning up the unconscious, what then is the function of deep sleep, apart from regenerating the body? Eckhart Tolle, bestselling author of *The Power of Now*, writes:

> *You take a journey into the Unmanifested every night when you enter the phase of deep dreamless sleep. You merge with the source. You draw from it the vital energy that sustains you for a while when you return to the manifested, the world of separate forms. This energy is much more vital than food.*
>
> *(Tolle 1999: 110)*

It is important to understand the nature of sleep and dreams. They are the prototypes of altered states of consciousness. I believe that most, if not all, kinds of altered states of consciousness are akin to the dream state. Thus, when we are in deep meditation or have taken a mind-expanding substance such as ayahuasca, we enter a state somewhat similar to dreaming. In Tibetan Yoga it is a declared goal to remain aware during dreaming and even during deep sleep. So we could say that when we enter an altered state through artificial means (meditation, hypnosis or drinking ayahuasca), we try to move consciously into sleep and dream states and thus confront the unconscious.

In my own research into altered states of consciousness it was important for me to understand how the brain and the brain waves related to different states or levels of consciousness. It was also essential for me to study changes in the balance of brain activity between the two hemispheres during an altered state. Here are some of my research results regarding meditation, 'transfer' of energy, feeling release therapy and the ritual use of ayahuasca.

MEDITATION: AIMING FOR A HIGHER CONSCIOUSNESS

Meditation has been known in the East for thousands of years. It is very widespread there, especially in India. In the West it has become popular over the last few decades and has attracted some interest from parts of the scientific community. Fifty years ago Bagchi and Wenger found increased EEG alpha activity in the brain during meditation (Bagchi and Wenger 1957) and since then many studies have confirmed that finding. It has also been established that if you meditate very deeply, the brain starts producing slow theta waves, signifying access to subconscious, emotional and dreamy states.

About the same time as Bagchi and Wenger published their study, two Frenchmen, Das and Gastaut, published the results of a study where they had found increased alpha waves during meditation substituted by fast beta and gamma waves, up to 40 Hz, during what they called *ecstasy* (Das and Gastaut 1955). This study was the first to point out the occurrence of 40 Hz gamma waves during meditation, even though the significance of these waves was not quite understood.

More recently US professor Richard Davidson and his group at the University of Wisconsin studied a group of Tibetan monks during meditation. Tibet's Dalai Lama sent eight of his most accomplished meditators to Davidson's laboratory for the study. Using quantitative EEG, Davidson's most important finding was a strong increase of gamma activity in the frontal cortex during meditation. (Lutz, Davidson *et al.* 2004)

Classical meditation as it is practised in the Hindu tradition in India takes place with closed eyes and with the attention directed inwards towards the body, for example focusing on the breath or on a mantra. This is designed to calm the thinking mind. During this form of meditation most studies have found increased alpha activity, especially in the posterior part of the brain.

In the Buddhist Zen tradition, meditation is often done with open or half-open eyes. Instead of focusing on the body, the meditator practises what is sometimes called *objectless meditation*, where they concentrate on love, compassion or thankfulness. This was the kind of meditation Richard Davidson was studying when he found strongly increased gamma activity in the brain.

A Kriya Yoga Study

Ancient Tantric Kriya Yoga is said to be a powerful method for awakening and expanding consciousness, as well as for strengthening vital and psychic energy. In conjunction with the Scandinavian Yoga and Meditation School and its leader, Swami Janakananda (a student of Indian Swami Satyananda), I conducted a number of EEG studies on yoga teachers with many years' experience of Kriya Yoga meditation (Hoffmann 1998).

Following two hours of Kriya meditation, the subjects' EEG measurements showed significant increases of alpha and theta rhythms (*see www.newbrainneworld.com*). During meditation, both alpha and theta wave activity often doubled or tripled, and even after meditation alpha activity stayed at a very high level. The increase of these rhythms was greatest in the rear part of the brain, where both alpha and theta rhythms rose by an average of 40 per cent after meditation.

There was a general tendency for the alpha rhythms to spread from the rear part of the brain towards the temporal and the frontal regions. In almost all subjects, alpha activity increased more on the right than on the left side of the brain, which we considered to reflect a greater involvement of the right hemisphere relative to the left during meditation. After the meditation all the subjects declared that they felt rested, focused and well.

The shift of alpha activity towards the right hemisphere seems to be a significant finding. Richard Davidson has shown that pessimistic, introverted or depressed people have more alpha activity in the left hemisphere (frontal and temporal regions), while optimistic, extrovert people have more on the right side (Wheeler, Davidson and Tomarken 1993). Thus, an increase of alpha activity on the right side is in theory a healthy change that may counteract stress and depression.

As we have seen, most EEG studies have found increased alpha and theta activity during deep meditation. There is little doubt that during these so-called deep states the subject has more conscious access to their lower levels of consciousness and is confronted by their own subconscious or unconscious mind. This process is associated with more alpha and theta waves. During deep meditation more and more neurons in the brain are recruited and work in synchrony at a slower and slower pace, building up the amplitudes (voltage) of alpha and theta waves. I suggest this is an attempt by the brain to confront, contain and eventually integrate unresolved emotional stress and conflict in the unconscious.

The growing interest in meditation in the West is probably due to the fact that more people are getting more stressed and are getting all kinds of symptoms as a result. However, if you are very tense and try to meditate, quite often you will not succeed. Osho said that Westerners were not able to meditate because they were too tense. In consequence of this view, he introduced his 'dynamic' meditation where you do some vigorous physical exercise, for example dancing, jumping and screaming, before you meditate. During this heavy physical activity, tensions are dissipated, which will make it easier to relax and meditate. Osho says, 'The more strongly you exercise physically, the deeper you go in meditation afterwards.'

During deep meditation it is very important that alpha activity dominates theta activity most of the time. The moment theta waves become stronger than alpha ones, you become unconscious. That is hardly meditation! So, when you meditate, there are at least two pitfalls to avoid.

One is that you start thinking, which will suppress alpha activity; the other is that you get drowsy, which will also decrease the alpha waves. If you can maintain a state of relaxation where your mind is calm but still very alert and focused on the inner world, then you are meditating in the right way and your alpha activity will be at a maximum.

Deeksha/Shaktipat: *The 'Transfer' of Energy*

In the spring of 2005 I came across a phenomenon called *deeksha*, or Oneness Blessing, an old tradition that has been further developed by Indian gurus Amma and Bhagavan Kalki, founders of the Oneness Movement in southern India.

Deeksha, sometimes called *shaktipat*, is a form of blessing where an initiated person puts their hands on another person's head for a few minutes with the intention of 'transferring' energy to that person. In the ancient Hindi scriptures, it is described as divine energy being 'transmitted' from an enlightened master to his devotees in order to stimulate their growth toward a higher consciousness. I put the word 'transmitted' in quotation marks since I believe that energy is not really transmitted. Rather I see *deeksha* as a phenomenon in which high consciousness vibrations in a *deeksha* giver, through resonance, trigger the same kind of consciousness vibrations in the *deeksha* receiver.

At the Mental Fitness and Research Centre in Copenhagen I had the opportunity to do a few EEG studies of people from Denmark and Sweden who had been initiated into giving *deeksha* in India. These studies were inconclusive, but nevertheless we found interesting brainwave changes in these people. There was a tendency in some (but not all) *deeksha* givers to increase their fast EEG activity (beta/gamma waves) for brief periods of time while they were giving *deeksha*. In the receiver of the *deeksha*, we most often saw increased alpha activity afterwards (*see www.newbrainnewworld.com*). Sometimes it even doubled.

These changes may have a natural explanation. When the *deeksha* giver concentrates on transmitting or channelling energy, it is only natural that their frontal lobes are activated and generate more beta/gamma waves. Also, the receiver of the *deeksha*, who is expecting to feel more energy, relaxes, opens up and perhaps therefore produces more alpha activity. A study including a control group is needed here to see if 'pseudo-*deeksha*' from an untrained person has the same effect as *deeksha* from an 'authorized' *deeksha* giver.

Our preliminary findings in Copenhagen were so interesting that we wanted to conduct a more rigorous study at the Oneness University in India. We agreed with the *dasas* (Indian guides) that we should go to the Golden City ashram and attend a 21-day process starting on 6 January 2006.

The 21-day process at the Oneness University consists of daily hour-long meditations plus guidance from the *dasas*. It also includes a number of rituals (*homas*) around a fire, accompanied by mantras, as well as rituals around silver sandals called *sadhukas*. *Deeksha* is given most of the time in the meditation hall by the *dasas*. Sometimes it is hands on, sometimes not.

Prior to the study, we gathered a research team in Denmark and Sweden consisting of my partner Inger Spindler, a neurotherapist from Copenhagen, Harald Kjellin, a professor at Stockholm University, and myself. A research protocol was written describing the methods for an EEG brain-mapping study of 12 subjects before and after the 21-day course.

Since it was not possible to establish an appropriate control group, we decided that our experimental group should act as its own control, since we took our own measurements both before and after the process. We also included psychological interviews and testing of all the subjects in order to evaluate possible psychological changes.

Quantitative EEG and brain maps were taken and psychological testing was done on 12 voluntary participants before and after the process. We analysed 26 EEG variables, but only found statistically significant changes in a few of these after the process. It came as a surprise that there were no significant change in alpha or theta activity after the course. Some subjects increased and some decreased their alpha activity, possibly because they were at different stages of consciousness development.

A solid finding, however, was a strong tendency for the two brain hemispheres to function more symmetrically following the process. We found this in 11 out of the 12 subjects. This balancing of the left and the right brain was, however, only present in the posterior (occipital-parietal) areas. In many cases we also found a distinct shift of brain activity from the left to the right hemisphere.

Another interesting finding was an increase of fast gamma wave activity in the frontal lobes after the process. Since these measurements demanded a special technical procedure, we were only able to measure 6 out of the 12 subjects. However, all 6 subjects showed substantial increases in gamma activity in the prefrontal cortex following the process. The group mean gamma amplitudes increased about 50 per cent in the prefrontal area.

With this small group of 12 people, we did not find any significant correlation between the scores from the psychological questionnaire and the EEG data, except for one important variable: the two people who benefited the most from the 21-day process, according to the psychological evaluation, both had exceptionally high frontal gamma levels. They were both rated to be in a high state of consciousness.

These were interesting initial findings. However, more studies need to be done in order to reach a safe conclusion about the effects of *deeksha*.

According to Bhagavan, one of the recent developers of the process, the *deeksha* energy affects the brain, the spinal cord and the chakras in certain ways. He says that activity in the rear part of the brain (the parietal lobes) is reduced, while the frontal lobes are activated. This change of dominant activity from the rear to the front of the brain (which we also see during certain types of meditation) is in my view significant in terms of the future evolution of the brain.

A Case of Awakening

I was fortunate to have the opportunity to draw up a brain map for a 60-year-old woman, Nandan, who had completed the 21-day course in Golden City. Four months prior to the retreat, she had lost her daughter, who had died during pregnancy, apparently due to a misdiagnosis. She writes:

> *This shock brought me so much grief, pain, anger and despair that I didn't know how to go on with my life anymore. Everything just felt dull and without any meaning. Then I suddenly remembered my daughter Disha telling me about the 21-day* deeksha *retreat in the south of India and that she was very drawn to do it. I felt very strongly that this was what I needed to do and so my husband booked it for us immediately.*
>
> *In the retreat, the first thing that happened was that all the pain of having lost my only child came up again. It felt like an unending flood of tears bursting out of me, first with a lot of pain but then, as I stayed with it and faced it fully, it started to transform into a strong longing for God. And then I suddenly felt a deep presence, as if everything around me was stopping and there was only a deep space inside me, and I recognized that in this space Disha was always present. It felt like finding a new way of being with her and at the same time also finding my real Self. Now, by giving more time to meditation and going inward, I am learning to focus more on recognizing her presence than on mourning her absence.*

In this way the 21-day deeksha *retreat was an incredible gift and healing process for me. Yes, I still feel pain, and probably always will, but the pain has a sweetness now through the growing acceptance of what happened and what is. The desperateness is gone and my heart is opening again, especially through feeling a deep connection with the Divine which is always present. We just need to feel that presence, give it space and there it is…*

It was quite amazing for me to be able to see this transformation process reflected in [my] brain-map pictures. It's incredible. Thank you.

(Nandan, Italy, 2006)

From Nandan's description of her experience and my general impression of her, I was led to believe that she was in an awakened state. Considering the enormous pain she had been through, it is amazing that her baseline brain map, when relaxing with closed eyes, was completely normal, with absolutely no signs of stress. When she was gazing at a picture of Bhagavan, which was of high significance to her, a large amount of fast EEG activity (beta and gamma waves) took place in her frontal lobes, which was most unusual. Since more gamma appeared on the left than on the right side, the likely interpretation (according to neurophysiologist Richard Davidson) is that she was in a stress-free, optimistic state of mind (*see newbrainnewworld.com*). This could very well be the EEG signature of an awakened consciousness.

PRIMAL THERAPY: RELEASING EMOTIONAL BLOCKS

In order to attain a higher state of consciousness, it's important that the brain and nervous system are not emotionally blocked. Meditation can increase awareness, sensitivity and access to lower levels of consciousness, and possibly dissolve certain blocks. However, sometimes stronger methods are called for. In feeling release (primal) therapy, blocked emotions are released through the verbal and non-verbal expression of the underlying feelings and traumas that gave rise to them. This may have a profound effect on the brain and nervous system.

In the mid-nineties I spent two years doing research at the Primal Center in Los Angeles, California, with Dr Arthur Janov, the founder of primal therapy. During a therapy session the patient is encouraged to freely express their feelings about any subject. Usually, they start talking about their present problems and feelings. Then, maybe suddenly, they access older repressed feelings that are similar in nature to what they are talking about. When that happens, the therapist encourages them to 'stay with the feeling' or 'drop into the feeling'. If the patient does so, they usually experience emotional pain, screaming or crying bitterly while perhaps reliving a traumatic childhood scene. The purpose of the therapy is to express these feelings and thus release emotional blocks and consciously integrate traumatic experiences from the past. Following a therapy session, a patient often appears much more relaxed, spontaneous and optimistic.

During my stay in Los Angeles I had the opportunity to carry out an EEG study in co-operation with Arthur Janov. We wanted to see if there were any changes in brain function as a result of primal therapy. Twenty volunteer subjects with emotional problems were studied via quantitative EEG methods and brain mapping before and after one year of primal therapy. Another, smaller group, functioning as a kind of control group, was studied before and after three weeks of intensive therapy.

Interestingly, the subjective changes in the patients were shown to correspond to significant changes in brainwave activity. The most outstanding finding was increased EEG alpha baseline activity in the brains of the patients following the year of therapy. An increase in alpha waves means better mental relaxation and better conscious access to one's body and feelings. I have also come to believe that it reflects conscious integration of blocked feelings. Possibly it also indicates a better integration between the left and the right hemisphere.

The group mean alpha activity in the back of the brain (occipital, parietal and temporal areas) increased significantly and almost doubled for the whole group of patients (*see newbrainnnewworld.com*). Theta and beta amplitudes also increased significantly in some areas. In the frontal lobes, however, there were only small and insignificant changes in brainwave activity.

In the three-week control group we saw the same kind of changes, though smaller in magnitude.

There were also some interesting changes in the balance of activity between the two cerebral hemispheres after one year of therapy. Most of the patients initially had less alpha activity in the right than in the left hemisphere. This, I suggest, was due to emotional stress primarily affecting the right hemisphere. Following therapy, however, there were much larger increases of alpha activity in the right than in the left hemisphere. I suggest that as the patients solved their emotional problems during therapy, the pressure on the right hemisphere diminished and so alpha waves could start building up again in that hemisphere, indicating better relaxation and integration of feelings.

As the patients progressed in therapy, we saw a clear tendency on the brain maps for alpha waves to spread from the back to the front of the brain (this was also the case during meditation). This could reflect better conscious integration of feelings.

After therapy many patients felt that they had become more 'real', closer to their real selves. Many described being 'thrown into themselves' during a therapy session. And in fact they were: they were thrown into their right hemisphere and even into the deeper parts of their brain.

But how do we explain the difference between the 'real' and the 'unreal' self? Janov (1996) says that repressed pain splits the self into two warring sides. One is the 'real self', full of repressed needs and pain, the other is the 'unreal self', which tries to cover up the unmet needs with neurotic behaviour such as obsessions and addictions.

The most fascinating finding in the primal study was when we did real-time brain mapping of patients during therapy. One patient was particularly interesting. After talking to the therapist for ten minutes about her dreadful childhood experiences, she started to relive a traumatic scene from her childhood. At the exact moment she dropped into the feeling, high-amplitude delta and theta waves started shooting up in various parts of her brain – predominantly 4 Hz delta/theta waves with amplitudes of several hundred microvolts appeared practically all over the brain. This unusual EEG pattern was repeated several times and was highly correlated with her emotional behaviour, which was monitored by a video camera placed in the therapy room.

I was astounded when I saw this. I realized that at that moment the patient was in the grip of an ancient brain, yet nevertheless able to consciously observe what was happening (*see newbrainnewworld.com*). This showed that it is possible under certain circumstances to have direct conscious access to a deep unconscious level characterized by theta and delta brain waves. I felt this was a breakthrough in my research – an exciting moment.

How the Cerebral Hemispheres Handle Emotions

How does the brain deal with emotions, whether traumatic or otherwise?

Emotions originate in the limbic system and will initially and primarily affect the right brain. The left hemisphere, the primary carrier of human consciousness, is not equipped to handle them. In fact I believe it often resents them and tries to avoid them. It seems to be able to block them in a number of ways, for instance through incessant thinking or speaking. Such over-activity in the left brain tends to suppress right-brain activity via the *corpus callosum* and yet the right brain is much better equipped to handle emotions than the left brain. In his book *The Right Brain and the Unconscious*, Rhawn Joseph writes:

> *Whereas these [limbic unconscious] impulses are almost completely foreign to the language-dependent conscious mind, and the left half of the brain, the right brain, being more involved in emotional functioning, is often (but not always) able to discern and recognize these limbically induced feeling states and desires for what they are.*
>
> *(Joseph 1992: 119)*

When an emotion arises and is expressed in its totality, it will usually fade away quickly. However, if it is suppressed or only partly expressed, it will leave a tension in the system, which often makes the person start thinking about the incident that provoked the emotion. Such thinking can prolong and build up very destructive feelings.

A Harvard psychologist, Julian Jaynes, who developed the theory of the 'bicameral mind', where one part of the brain 'speaks' while the other listens and obeys, proposed a two-tiered theory of emotion differentiating between affect and *feeling*. An affect is the immediate psycho-physiological response to a stimulus and the associated subjective experience. Fear, for example, is the response to a threatening stimulus. When the stimulus is gone, the psycho-physiological response dies out quickly, as does the subjective experience – unless you give it time. If time is added, so is the ability to remember and anticipate, and that can make the affect persist.

Since only the left hemisphere is preoccupied with time (with remembrance and anticipation), only the left hemisphere has the ability to perpetuate the affect (for example, fear) and turn it into a feeling (for example, anxiety).

Therefore, while the right hemisphere seems to be able to let go of these responses very quickly, the left hemisphere has the ability to perpetuate them as feelings, in theory infinitely. A person who had a frightful experience in the past and therefore keeps anticipating a similar experience in the future might suffer from chronic anxiety, for example. Anxiety is always anticipatory.

Julian Jaynes suggested that in ancient (pre-bicameral) times, people predominantly used their right hemisphere. They had no feelings, only short-lived affective reactions. It was only much later, when the left hemisphere became dominant through the development of language and thinking, that they were able to have extended affective reactions. It was then that fear turned into anxiety, aggression turned into anger and hate, mating became sex, shame turned into guilt and pain into suffering. If this theory holds true, destructive feelings such as anxiety, anger and hate (and their companions, envy, jealousy and vengefulness) belong to the domain of the left hemisphere. If we want to liberate ourselves from these feelings, we have to pacify the ego and move from the left to the right hemisphere.

I want to emphasize that there is nothing wrong with short-term reactions such as aggression in appropriate situations. However, if, through thinking and anticipation, you perpetuate aggression and turn it into anger and hate, it becomes very destructive.

If you really want to find out how to manage destructive feelings, it might be useful to explore ways of becoming more aware of the right hemisphere's mind. One safe way of doing this would be meditation or some other kind of awareness training.

Jill Taylor wrote that from her right-hemisphere perspective she resented the return of her old emotional self as her left hemisphere started recovering. She discovered, however, that if she was feeling angry, she could choose to move into the present moment (right hemisphere) 'allowing that reaction to melt away as fleeting physiology': 'I learned that I had the power to choose whether to hook into a feeling and prolong its presence in my body, or just let it quickly flow right out of me.' (Taylor 2008: 120)

Feeling Release versus Presence

It is important to clean both body and mind of impurities in order to prepare ourselves for higher states of consciousness. Arthur Janov coined the phrase *primal pain*, which he sees as an accumulation of repressed feelings, mostly from childhood, residing in the unconscious. Through the expression of these repressed feelings in primal therapy, as we have seen, the traumatic experiences can be integrated and the pain dissolved.

Eckhart Tolle also talks about accumulated pain from the past, which resides deep in the unconscious. He calls this the *pain-body* and says that it has two modes of being, either dormant or active. In some people it is dormant most of the time, while in others it is active most of the time and can be triggered by even the most trivial external event:

> *The pain-body consists of trapped life-energy that has split off from your total energy field and has temporarily become autonomous through the unnatural process of mind identification.*
>
> *(Tolle 1999: 32)*

In Tolle's view, basically all emotions are modifications of one original undifferentiated emotion, which he calls *pain*, which has its origin in the loss of awareness of who we are and our disconnection from our source, or God. This basic pain seems to be the karma of the whole human race. On top of that, we accumulate painful childhood experiences and traumas and it all adds up to what Tolle calls the pain-body. I believe this is somewhat similar to what Janov calls primal pain, except that Janov only recognizes pain suffered in this lifetime.

The two authors suggest different strategies to handle the pain. Janov says that the painful feelings have to be felt and integrated, and admits that it may take years of feeling therapy. Tolle, on the other hand, says that whatever you need to know about your unconscious past will be brought out by the challenges of the present. If you delve into the past, as you do in feeling therapy, where you often have to feel the same feeling over and over, you will eventually realize that it is a bottomless pit since there is always more. If you can give full attention to your behaviour in the present, to your thoughts and feelings, you will see the past in you.

Tolle says:

> *If you can be present enough to watch all those things, not critically or analytically but non-judgmentally, then you are dealing with the past and dissolving it through the power of your presence.* You cannot find yourself by going into the past. You find yourself by coming into the present *[my emphasis].*
>
> *(Ibid: 75)*

Both Tolle and Janov say that it is important that you do not try to resist the painful feelings ('what you resist persists'). The pain should be confronted, observed and felt. Sometimes you may have to express it as in feeling therapy, but you shouldn't create a script in your mind around it, such as 'I am a victim of incest' or 'I was an abused child.' So the way I see it, the recipe is: confront the painful feeling, accept it is there for now and go deep into it while remaining intensely alert – feel the energy of it without thinking.

This kind of 'awareness therapy' may eventually integrate and dissolve the pain, even if it takes time. Such an approach would be totally in line with Eckhart Tolle's advice as well as with Osho's view that 'Awareness is enough.' Sometimes, though, there may be so many heavily repressed feelings that you need to express them through some kind of feeling therapy.

While Janov believes that connecting a repressed feeling with a past event is crucial, I feel that isn't so important. It is the spontaneous abreaction that has an effect. By analysing it and trying to understand the cause of the feelings, you will just be digging a bigger and bigger hole and you will never get to the bottom of it. It is true that sometimes during or after a therapy session you may get a spontaneous insight into the connection between a feeling and an old event. That's OK, but it should not be analysed. Also, remember that the more attention you give to the past, the more you energize it.

In my view, the beneficial effect of feeling therapy is primarily due to the physical/emotional release of blocked or repressed feelings. These constitute a barrier that prevents you from accessing your deeper levels of being. Through feeling therapy it is possible to break through this emotional barrier and connect to the deepest levels of being. However, feeling therapy should always be followed up by meditation or awareness training (for example brainwave training) in order to focus the released energy consciously. The combination of the two methods, I feel, will constitute a very powerful tool in consciousness development.

AYAHUASCA: A JOURNEY INTO THE UNCONSCIOUS

Ayahuasca is a bitter drink composed of extracts of plants found in the rainforest around the Amazon delta in South America. It includes the leaves of the plant *Psychotria viridis*, which contain the psychoactive drug dimethyltryptamine (DMT). This has a structural resemblance to serotonin, which is a chemical transmitter naturally found in the brain. When DMT binds to the nerve receptors in the brain, a change in consciousness takes place.

Native South American Indians and shamans have used ayahuasca for at least a thousand years for spiritual and healing purposes. Its ritual use for spiritual purposes is completely legal in Brazil. Today large religious organizations such as Santo Daime and Uniao do Vegetal in Brazil use it on a regular basis as a holy drink at their spiritual ceremonies. Unlike a chemically synthesized hallucinogen such as LSD, with all its unknown risks, ayahuasca is prepared from natural plants and no harmful side-effects, to my knowledge, have been reported over the last thousand years.

Only recently has Western science shown an interest in ayahuasca. In 1993 American and Brazilian scientists conducted the Hoasca Project in the Amazon Delta, where a group of long-term users of ayahuasca (belonging to the religious organization Uniao do Vegetal) was compared to a group of non-users (Metzner 1999). In addition to biochemical, pharmacological and physiological investigations, the subjects were given a thorough psychological/psychiatric examination. A large percentage of the long-term users of ayahuasca were cured of alcohol and substance abuse. Compared to non-users, they were found to be more trustworthy, loyal, optimistic, energetic and emotionally mature. In addition, the long-term users achieved better results in tests measuring concentration and short-term memory. No side-effects were found (Grob, 1999).

An EEG Study of Ayahuasca in a Ritual Setting

In the spring of 2000 I went to Brazil to do an EEG field study of a group of people participating in a workshop in which ayahuasca was consumed. The study was undertaken in co-operation with a Dutch medical doctor, Michael Hessellink, and the Brazilian shaman, Yatra da Silveira Barbosa, who conducted the workshop. It took place near a small town called Alto Paraiso, a few hours' drive north of the capital, Brasilia, in a few small primitive huts and an open meditation hall called the Zendo in the middle of the jungle. There were no baths, no real toilets and no electricity. Thus I had to bring several extra batteries for my laptop.

During the one-week workshop, four ayahuasca rituals were held in the Zendo. Three rituals were held from 6 p.m. until 12 a.m., and one was held early in the morning before sunrise. The Zendo was decorated with flowers and candles. Everybody taking part in the ritual dressed in white clothes and sat in a circle on mattresses or on small pillows on the floor. Much of the time was spent meditating while listening to recorded spiritual music. At times someone played a guitar and Yatra sang beautiful songs. It was a wonderful ceremony.

Three times during the ritual people lined up to get 75 ml of ayahuasca tea. As the evening progressed, people were obviously getting high and going into altered states of consciousness. Some were sitting or lying down, completely absorbed in deep meditation, while others were dancing and laughing. Everybody appeared to be having a good time.

Twelve participants volunteered for the study. During two of the evening rituals their EEG recordings were taken at the end of the ritual, around midnight. At this point people were still very high, but some were slowly coming down to normal reality again. The recordings were taken while the subjects were still lying down on their mattresses. They were awake and able to communicate in a normal way.

The results of the study showed that at the peak of the ayahuasca experience there was a strong and significant increase in both alpha and theta waves in most parts of the brain relative to the normal resting state. Beta wave activity, however, was unchanged. The largest increases in alpha activity were observed at the back of the brain, but the alpha activity of the frontal lobes remained unchanged. The theta waves, on the other hand, increased in many parts of the brain, including the frontal areas.

Increased theta wave activity in the frontal lobes will strongly hamper their function and weaken attention and control. There is a risk that the subject will be flooded with unconscious material that cannot be controlled by the frontal cortex. This underscores how important it is to stay alert and observant during the ayahuasca experience. Steady, unwavering attention to what is going on in the mind, without any attempt to avoid or suppress anything, is the best protection against being overpowered by the strange dark forces of the unconscious.

There is no doubt that the increased EEG alpha and theta activity after drinking ayhuasca reflected an altered state of consciousness. The subjects reported increased awareness of their subconscious processes, which included lots of dreamlike imagery. The altered state of consciousness induced by ayahuasca has some similarities with meditation, and ayahuasca may be regarded as a magnifier of meditation, since it helps pacify the ego and increase access to the unconscious.

The results of the study showed that at the peak of the ayahuasca high there was an enormous increase in both alpha and theta waves, especially in the posterior region of the cortex and on the left side (*see www.newbrainnewworld.com*). Alpha amplitudes in the left hemisphere increased much more than in the right in 10 out of the 12 subjects, in some by as much as 200 per cent. I believe that ayahuasca strongly activates the limbic system and the right hemisphere, which then overpowers the left hemisphere through inhibition via the *corpus callosum*. Before drinking ayahuasca, all the subjects except one had dominant left-hemisphere activity, and after drinking the tea, they all had dominant right-hemisphere activity. (This was indicated by a reversal of their inter-hemispheric amplitude relationship. Such a reversal was also found by Goldstein and Stoltzfus in their study of subjects under the influence of psilocybin [Goldstein and Stoltzfus 1973].)

Apparently ayahuasca shuts down the left hemisphere, so that right-hemisphere activity prevails, and simultaneously hampers the frontal lobes. It looks as if it turns the left hemisphere into a passive observer of what is going on in the right hemisphere and the unconscious.

These physiological changes are in accordance with the reported subjective changes. Many subjects felt that after drinking ayahuasca they were pushed into a dream world while wide awake. Some felt that their egos were crumbling, or even dying. This supports the hypothesis that the left hemisphere is the seat of the ego.

Following an adequate intake of ayahuasca, I believe it is possible to go fully conscious into the dream state. The process seems to be accompanied by a lowering of the brainwave frequency (more alpha/theta waves), especially in the frontal lobes and in the left hemisphere. During this process, dissolution of the ego – also called ego death – will often be experienced. This can be a very scary experience. But if you open your eyes and start talking or doing something, this will activate the left hemisphere and the frontal cortex and immediately pull you (more or less) out of the altered state. This is an important reason why it is much safer to take ayahuasca than LSD, where you can get caught up in very scary states without a chance of getting out.

It is also interesting to note that many people have lucid dreams for several nights after drinking ayahuasca.

An Interview with Yatra Barbosa on Ayahuasca

Ayahuasca, meditation and feeling release therapy all affect right-hemisphere activity and facilitate access to the unconscious. All three methods tend to increase both alpha and theta waves in many areas of the brain. The ritual use of ayhuasca has now become a profound tool in feeling release therapy and energy work. Yatra da Silveira Barbosa, who ran the workshop where I conducted the study, has developed a unique therapy combining the use of ayahuasca with meditation and feeling release therapy. The primary goal of her workshops is to promote spiritual growth and the transformation of consciousness. According to Yatra, meditation and feeling release therapy are tools which we can use, with ayhuasca as a catalyst, to break through emotional blocks to deep unconscious levels, thus paving the road to higher consciousness.

Yatra was born in Brazil, where she studied the use of psychedelic plants with native shamans. For many years she also underwent training at Osho's School of Mysticism in Poona, India. She lived for some years in Amsterdam, Holland, where she founded the organization Friends of the Forest to preserve the rainforest and native tribes in Brazil, where she is now living again. I interviewed her in Amsterdam after a workshop in 1999:

Q: **What are the most obvious subjective changes of consciousness following the intake of ayahuasca?**

A: *The possibility of contemplating the Self without the interference of the analytical and critical mind, therefore the possibility of accepting and integrating the Self at a higher level of consciousness.*

Q: **How would you describe your state of consciousness following the intake of ayahuasca? I remember that you once claimed to be in 'a state of total awareness'. What did you mean by that?**

A: *I meant not thinking, not judging, just witnessing and being. Of course that is my own experience. I cannot say that it will be like that for everybody. Your experience depends on the level of consciousness you are at.*

Q: **Is this state of 'total awareness' a high-energy lucid state of consciousness? And how would you compare it to the 'twilight' or hypnagogic state between waking and sleeping?**

A: *It is a higher octave of energy, since you are totally awake in a lucid state of consciousness. There is no comparison with the 'twilight state'. Here again, I can only speak for myself – I cannot say it is like that for everybody.*

Q: Why do you think that ayahuasca tends to induce lucid dreams?

A: *It is because you have awakened the connection to your unconscious. You can even watch your dreams consciously. That happens to me very often. Sometimes I wake up in the middle of a dream and then go back to sleep to continue the dream in order to see the outcome. Although I cannot alter its course, I can watch it.*

Q: What is in your mind the importance of cleansing the body physiologically, through diet, and psychologically, through feeling release and meditation?

A: *The cleansing is very important physiologically and psychologically, otherwise there is no room for ascendance of the energy, since there will be physiological and psychological blocks. For the energy to run freely in the physical and the subtle bodies, the blockages have to be dealt with and dissolved. Then the chakras will be aligned and the energy able to flow.*

Q: How do you evaluate modern psychotherapy using primarily feeling release and talk therapy, leaving out spirituality?

A: *Talk therapy keeps the person in the mind, as it is done through intellectual impressions. Only primary feeling release therapy, with emphasis on spirituality, will bring the person in touch with their true feelings. This will open their heart space to love and open doors to higher levels of consciousness. In the first place, man is a spiritual being, and one way or another, he carries a spiritual search within himself, which is the reference point to who he is beyond his physical body.*

Q: What do you mean by 'breaking through the emotional body to love, self and God'?

A: *When we break through the emotional body, we poke the veil that is keeping us away from experiencing our true feelings and our true Self. Breaking through this veil brings us back to our inner Self without separation, to our inner truth, to our being of pure love and joy, to the divine within. That is God within us.*

Q: What are the higher spiritual purposes of ayahuasca rituals?

A: *To heighten the level of consciousness and to make us aware that we are beings of light.*

Q: What does ayahuasca do in terms of love? Does it affect the heart chakra?

A: *Ayahuasca affects all the chakras, dissolving the blockages and creating unity.*

Q: Do you have any experience of adverse psychological effects of drinking ayahuasca?

A: *Yes. There are adverse psychological effects with higher doses of ayahuasca in people presenting certain disorders. We refuse to give ayahuasca to people with severe psychiatric disorders. Ayahuasca is not for everybody. Sometimes it can cause a manic state in people with psychological or mild psychiatric disorders, but they will go back to their normal behaviour by the end of the session.*

Q: Do you think that ayahuasca can heal mental disorders?

A: *Yes, many, but not all.*

Q: Does the regular use of ayahuasca facilitate meditation and creativity?

A: *Without doubt.*

Q: I have found in my research that ayahuasca and feeling release therapy create the same kind of EEG changes, namely increased alpha and theta activity. Do you think that ayahuasca could be used as an adjunct to psychotherapy? And how would that work?

A: *Yes, ayahusaca could be a great help to psychotherapy. They would work together and shorten the recovery time.*

Q: Can you explain why ayahuasca therapy would be especially good for treating drug addicts?

A: *The drug addict is a shaman using the wrong carburettor. They are on a spiritual search for a higher state beyond their emotions. Ayahuasca prepares them to deal with their issues by bringing them into their heart space, reconnecting them with their true feelings and inner truth. It also realizes their spiritual search, establishing their connection with their true Self and the God within them.*

THE SIGNIFICANCE OF THE RIGHT HEMISPHERE IN ALTERED STATES

I have no doubt that the right cerebral hemisphere and the limbic system play a major role in altered states of consciousness. The research I have done shows that the largest EEG changes take place in the right hemisphere following meditation, feeling release and the ingestion of ayahuasca.

In general we would expect to see a symmetrical distribution of electrical activity between the two hemispheres, reflecting an even balance of activity between them. However, a number of studies have shown that in normal control subjects, the amplitudes tend to be slightly higher in the right than in the left hemisphere. Mean alpha amplitude is often used as an indicator of cortical brain activity: the lower the alpha amplitude, the higher the brain's activity or arousal level. Other studies have shown that in patients with stress, alpha activity in the right hemisphere tends to be lower than on the left side. My own observations in hundreds of people have shown that most, if not all, of those with severe emotional stress have lower alpha amplitudes on the right side. This imbalance, however, is reversible and usually normalizes after successful treatment.

My own research, as well as Richard Davidson's, shows that hemispheric asymmetry of alpha amplitudes is much more critical in the front areas of the brain than the back.

Professor Richard Davidson of the University of Wisconsin has shown that people with lower alpha amplitudes on the right side tend to be more inhibited, negative and prone to depression than people with lower alpha on the left side, who, in contrast, are more outgoing, positive and optimistic. These asymmetries, according to Davidson, are already present to some extent at birth and seem to reflect differences in basic personality traits.

Asymmetry between the hemispheres of EEG amplitudes probably reflects how the two hemispheres interact with each other. In an outgoing, optimistic personality, alpha activity is lower on the left side, indicating that the left, verbal-analytic hemisphere is more active than the right hemisphere. At the same time, in this outgoing personality,

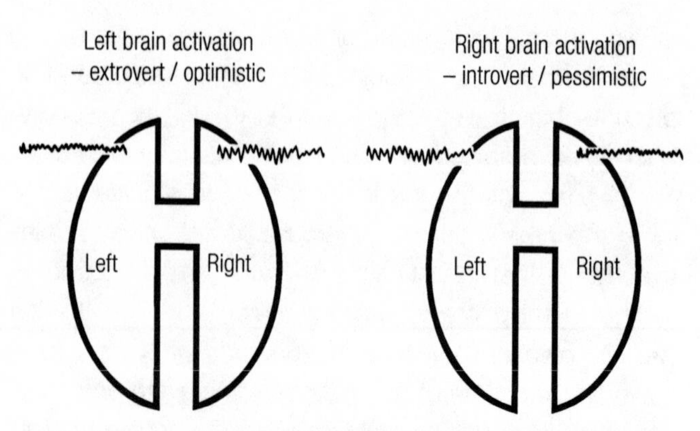

Figure 3.3: Asymmetry of Alpha Waves Reflects Personality Type

alpha waves are highest on the right side, possibly reflecting that the right hemisphere is better connected to the feeling parts of the brain and thus better able to handle emotional pressure. In the inhibited, pessimistic personality, alpha activity tends to be higher on the left side, which is also the case with people under stress. It seems that emotional pressure from lower levels of the brain (the limbic system) reduces alpha activity primarily in the right hemisphere, while at the same time, as a consequence of reciprocal inhibition between the hemispheres, it increases on the left side, possibly hampering the optimistic, outgoing qualities of the left hemisphere.

In connection with drinking ayahuasca or taking other mind-expanding drugs, one may speculate about the risks of one or the other hemisphere being overwhelmed by lower-level unconscious pain. According to neurophysiologist Rhawn Joseph, lower-level emotional pain is much better tolerated by the right feeling hemisphere than by the left verbal brain, which is unfamiliar with these 'upshooting strange forces' and unable to interpret them. (Joseph 1992) Thus, higher amplitudes in the right hemisphere, as seen in most people, probably indicate that this side of the brain is closer to and more integrated with feelings originating deeper in the brain.

If the rise of emotional pain interferes with left-hemisphere functioning, we may expect mental disturbance of a psychotic dimension, accompanied by larger amplitudes on the left side. This is exactly what several studies have shown. Goldstein and Sugerman (1969) found in a group of psychiatric patients lower amplitudes on the right side (right-hemisphere activation), while a group of normal controls showed lower amplitudes on the left side (left-hemisphere activation).

Interestingly, normal subjects under the influence of hallucinogens, as well as during dream (REM) sleep, show predominantly right-hemisphere activation. Thus, there seems to be a certain resemblance between some psychiatric patients on the one hand and normal people during naturally occurring or drug-induced hallucinatory states on the other hand. In both cases the right hemisphere seems to be highly activated.

A BASIC NEUROLOGICAL MODEL OF ALTERED STATES

What happens in the brain during altered states of consciousness when people are poking their unconscious? The results of my research show that such states, whether induced by meditation, reliving past traumatic experiences (as in primal therapy) or drinking ayahuasca tea, exhibit somewhat similar changes of brainwave activity. These changes are increased alpha and theta waves over the whole rear part of the brain, together with changes in the balance of activity between the left and the right hemispheres. Interestingly, the frontal cortex is the area least affected (on EEG criteria).

Most of these states have one thing in common: they give the individual better access to their unconscious, and this is reflected in the EEG by an increase of alpha and theta brain waves and a shift of brain activity toward the right hemisphere. Thus, in altered states, according to my model, the right hemisphere and the unconscious are activated while the frontal lobes seem unaffected or even more passive than normal. As we shall see later, this is contrary to a higher (awakened) state of consciousness, where primarily the frontal lobes are activated.

In my view, ayahuasca (and probably most other psychedelic substances) activates deep brain structures, primarily the limbic system and secondarily the right hemisphere. Left-brain activity is indirectly suppressed through reciprocal inhibition via the *corpus callosum* and the individual gains access to a whole new realm of subconscious and unconscious experiences through the right hemisphere.

There seems to be a dynamic mechanism in the brain causing increased activation of the right hemisphere at both low and high arousal levels in the brain and nervous system. Thus the left hemisphere tends to dominate at the medium arousal level seen during daily routines, while the right hemisphere usually dominates during very low arousal (for example meditation) and very high arousal (for example after drinking ayahuasca), and gives better access to the unconscious.

This was also the conclusion of Roland Fischer, professor at Maryland Psychiatric Research Center, after extensive studies of people who had taken the mind-expanding substance psilocybin:

> *Indeed, the model of Fischer and supporting data of Goldstein et al. and Goldstein and Stoltzfus indicate that a shift from left to right hemispheric information processing can be induced by levels of subcortical arousal which are above (hyper) or below (hypo) the levels of arousal associated with routine daily activities.*
>
> *(Fischer and Rhead 1974: 196)*

In summary, I propose that in most, if not all, altered states of consciousness, the left hemisphere, and to a certain degree the frontal cortex, is somewhat inhibited, while the right hemisphere and the limbic system are activated.

In his book *The Doors of Perception*, Aldous Huxley describes his experiments with mescaline, a natural psychedelic drug (from cactus) somewhat akin to ayahuasca. He theorizes that Consciousness is in fact unlimited, and in order to make biological survival possible, irrelevant information must be filtered out. Thus the function of the brain is to filter out everything other than consensual reality – infinite Consciousness has to be funnelled through a 'reduction valve' in the human brain. Huxley believed that mescaline simply put a crack in that mental filter. Maybe he had a point there, and that makes me ask: do mescaline, ayahuasca and other psychedelic drugs make a 'crack' in the *corpus callosum*, causing the bridge between the left and the right hemisphere to open up more?

Chapter 4

KUNDALINI: AN EVOLUTIONARY ENERGY IN MAN

Kundalini is a Sanskrit word meaning 'coiled'. It refers to a type of energy that is often symbolized by a coiled-up serpent resting at the base of the spine. The concept can be traced back to earlier than 3000 BC in various parts of the world. In India, it was mentioned in the famous Hindi scriptures the *Upanishads*. Much more recently in the West, the Swiss psychoanalyst Carl Jung took an interest in the phenomenon and tried to relate it to his own theories.

In most people this energy is inert or asleep, but it can be aroused spontaneously or through a number of practices such as yoga, meditation or *pranayama* (breath control) or through more dubious methods such as the intake of mind-expanding drugs such as LSD, psilocybin, mescaline and ayahuasca.

In her book *Kundalini and the Chakras*, Genevieve L. Paulson writes:

> *Involuntary ways in which* kundalini *may be released include drug use, overwork, a severe blow or injury to the tailbone area; grief, trauma or excessive fear; excesses in meditation, growth practices, or sex. Excessive sexual foreplay without orgasm may also cause spontaneous* kundalini *release.*
>
> *(Paulson 2002: 10)*

When the *kundalini* energy is aroused, it is said to ascend through the spinal column, passing through a number of energy centres called *chakras*. The ancient texts are not very clear about what exactly a chakra is, and modern science doesn't have a clue. One description comes from the Indian Tantric master Osho, who says that a chakra is kind of a wheel or a vortex, part of the subtle energy body, helping to propel the energy to higher levels. According to Osho, a chakra only starts working when energy flows into it. Then it helps the energy break through any blocks it meets on the way, so it can flow smoothly. Osho (1976: 81–108) and other Indian masters say that when the energy rises and activates the chakras, they act like switches that can turn on the corresponding brain areas. When the *kundalini* energy rises all the way to the top chakra (sahasrara), awakening is said to occur.

I must admit that theories of *kundalini* and the chakras lack scientific support and often appear somewhat speculative. However, these theories have persisted over thousands of years and are claimed to be facts by many masters in the

East. The problem is that they are based on these masters' personal experience. *Kundalini* energy cannot be measured directly by physical instruments. Indirect measures such as electrical skin potentials and EEG gamma waves have been used with some success. However, sooner or later, I feel sure that science will be able to obtain direct objective measures of this phenomenon.

Osho maintains that *prana* (life energy) flows through the *kundalini* circuit (the *nadis*) in a subtle, non-physical body called the etheric body, just as blood flows through the circulatory system in the physical body. He says the etheric body is found deep down in the physical body below the cellular level. Most texts claim that the *kundalini* circuit and the chakras are not located in the physical body, but the awakened master Gopi Krishna stresses that the nadis and chakras are physical, even though they are so fine as to be invisible to the naked eye.

I am sure that the chakras, the nadis and the subtle bodies are not immaterial. They are based on matter representing much higher frequencies than physical matter, and they are most likely located deep down in the body's microcosmic structure.

Swami Satyananda, founder of the Bihar School of Yoga in India and a student of the famous Swami Sivananda, wrote after his own *kundalini* awakening:

> *Kundalini is not a myth or an illusion. It is not a hypothesis or a hypnotic suggestion. Kundalini is a biological substance that exists within the framework of the body. Its awakening generates electrical*

> *impulses throughout the whole body and these impulses can be detected by modern scientific instruments and machines.*
>
> (Satyananda 1996: 4)

Ancient Hindi literature talks about three different channels through which *kundalini* energy can rise: a central channel, called Sushumna, and two side channels, called Ida and Pingala respectively. According to yogis, the energy is supposed to rise through Sushumna. However, sometimes it rises through either Ida or Pingala instead, creating a number of problems. Ida and Pingala are said to be related to the autonomic nervous system, which consists of a parasympathetic system (related to Ida) and a sympathetic system (related to Pingala). These two systems are in many respects antagonistic: the sympathetic system activates the brain, while the parasympathetic relaxes the brain.

In order for the energy to rise through the middle channel, Sushumna, there must be a proper balance of activity between the sympathetic and the parasympathetic systems. This balance is often disturbed, due to emotional conflicts, stress and psychiatric disorders. Yogis usually advocate *pranayama* (breath control) in order to stabilize the balance between the two systems.

Gopi Krishna wrote in his autobiography, *Living with Kundalini*, that in his case initially the energy was aroused through the sympathetic channel, Pingala, causing severe anxiety and even psychotic episodes. He believes that the human brain is still in a state of organic evolution towards a neurological basis for a higher level of Consciousness.

The driving force of that evolution, he claims, is *kundalini* energy:

> *The human reproductive system functions in two ways, both as an evolutionary and as a reproductive mechanism. As the evolutionary mechanism, it sends a fine stream of a very potent nerve-energy into the brain and another stream into the sexual region, the cause of reproduction.*
>
> *(Krishna 1975: 111)*

Gopi Krishna is convinced that *kundalini* is the cause of both madness and genius as well as the source of all creativity. He believes it is the most urgent subject for science to study:

> *The research that I am asking for will place in the hands of science the secret of genius, the secret of that Source from which science was born, from which philosophy, literature, music, from which all was born. The research will put science in touch with the source of all creativity, all nobility, and all psychic powers in man.*
>
> *(http://gopikrishna.us/articles/nobel.html)*

It would be in place here to warn against attempts to artificially arouse the *kundalini* energy prematurely. A premature *kundalini* awakening can cause severe problems, including psychosis, in people not ready for it. In my view, using synthetic LSD to activate the *kundalini* can be very harmful, while substances such as ayahuasca, peyote and mescaline probably are less hazardous (if you know what you are doing), since these substances are natural and have been used by indigenous people for hundreds of years.

American writer Darrel Irving (1995) describes in his book *Serpent of Fire* how he activated his own *kundalini* through the use of LSD, a method he most certainly does not recommend to others because of the hazards involved:

> *For the rush, the kundalini, no longer restrained, then rose swiftly, spiralling and whirling upward, gyroscoping up without impediment, up, up, up, expanding, expanding, expanding consciousness as it rose, continuing up the length of my spine and finally right out through the top of my head in an explosion of consciousness. The feeling was blissful beyond description, and suddenly, just like that, I was propelled to another state of consciousness all together. I found myself united, in a state of union, of oneness, with the universal Consciousness … with God.*
>
> *(Irving 1995: 53–4)*

Soon Irving was struck by the realization:

> *This is who I really am!!! For this was the real me. This was a new self, a complete self who superseded the old, shrunken, and only partial self I had previously perceived myself to be. For to be truly in touch with one's inner self, is not only to touch bliss, it is also to become a full person.*
>
> *(Ibid: 54)*

The reason why Irving here felt he was becoming a 'full person' with a 'new self' could be because the drug had activated his right hemisphere and opened up the connection between the hemispheres.

A PHYSIOLOGICAL MODEL OF *KUNDALINI*

Czech-born physicist Itzak Bentov has developed a brain model of the *kundalini* phenomenon, which has become quite popular among some researchers (Bentov 1977). I believe that his model is somewhat crude and mechanical. However, here is a short version of his theory: when the heart is pumping blood into the main artery, tiny mechanical vibrations (micromotions) develop in the upper body. If the body is completely immobile and respiration is slow and barely perceptible, as is the case in deep meditation, these micromotions will assume great regularity and oscillate at seven cycles per second (7 Hz). Subsequently, standing waves in and above the audio frequency range will develop in the ventricles of the brain. These may cause the high-pitched sounds meditators often hear, which may herald the coming of the *kundalini*.

Bentov further theorizes that the standing waves in the ventricles will stimulate an electrical loop circuit in the sensory cortex oscillating at 7 Hz (which incidentally is within the EEG theta frequency band). As this oscillating current passes through the sensory cortex in a closed circuit, it will stimulate different brain areas in sequence and create a number of bodily sensations, starting at the toes and moving up the body to the head. These may be the sensations that are reported to accompany *kundalini* awakening.

According to Bentov, the 7 Hz oscillating current in the sensory cortex also passes through the limbic system. Here it stimulates pleasure centres in the amygdala, thereby initiating the so-called *kindling phenomenon*. Repetitive stimulation (kindling) in this area may eventually lead to orgasmic activity. Animal experiments have shown that electrical stimulation of the amygdala for a few seconds each day for two or three weeks produces convulsions. The amygdala is the part of the brain most receptive to kindling; thus what Bentov calls the *Physio-kundalini Syndrome* seems to be based on physiological kindling in the same way that sexual orgasm is based on physiological kindling.

When *kundalini* energy starts moving in the body, lower brain mechanisms, such as the brainstem and the limbic system, are aroused. This may cause an imbalance of the autonomic nervous system, leading to either sympathetic or parasympathetic over-activity. Sympathetic over-activity may cause, for example, a panic attack, while parasympathetic over-activity may cause depression. This is why yogis emphasize the importance of having a stable autonomic nervous system when the *kundalini* becomes active. A quiet life, yoga and a proper diet are recommended in order to stabilize this system.

One popular exercise designed to create a proper balance between the sympathetic and the parasympathetic systems is *Nadi Shodan*, breathing alternately through the left and right nostril. In an EEG study I did in co-operation with the Scandinavian Yoga and Meditation School, I actually demonstrated that trained meditators were able to improve the balance of brainwave activity between the two hemispheres through a 20-minute exercise of alternate breathing through the nostrils (See *www.yogameditation.com*).

As mentioned earlier, there are no recognized scientific methods for measuring *kundalini* energy directly. However, a number of indirect measures have been used. Bentov used equipment for recording mechanical micromotions in the body and pulsating magnetic fields around the head, as well as electric skin potentials for tracing the movement of energy through the body.

In my own work I have used high-amplitude EEG gamma wave activity in the prefrontal cortex as an indicator of high-energy metabolism in this area possibly reflecting *kundalini* activity. At the Mental Fitness and Research Centre we had the opportunity of measuring the EEG of a young woman who claimed to be able to raise her *kundalini* energy on command. We recorded EEG gamma wave activity (30–42 Hz) from her left and right prefrontal areas while she tried to raise her *kundalini* energy (*see Figure 4.1*).

The *kundalini* energy seems to be reaching the brain's prefrontal areas approximately 12 seconds after the command was given and to be building up to a maximum in both hemispheres after about half a minute. After that, it drops down again. It's interesting to note that the left frontal area takes the lead, since it responds slightly faster than the right and climbs to a higher level. At the time of maximum amplitude (voltage), the phase synchronization of gamma waves between the left and right side is also

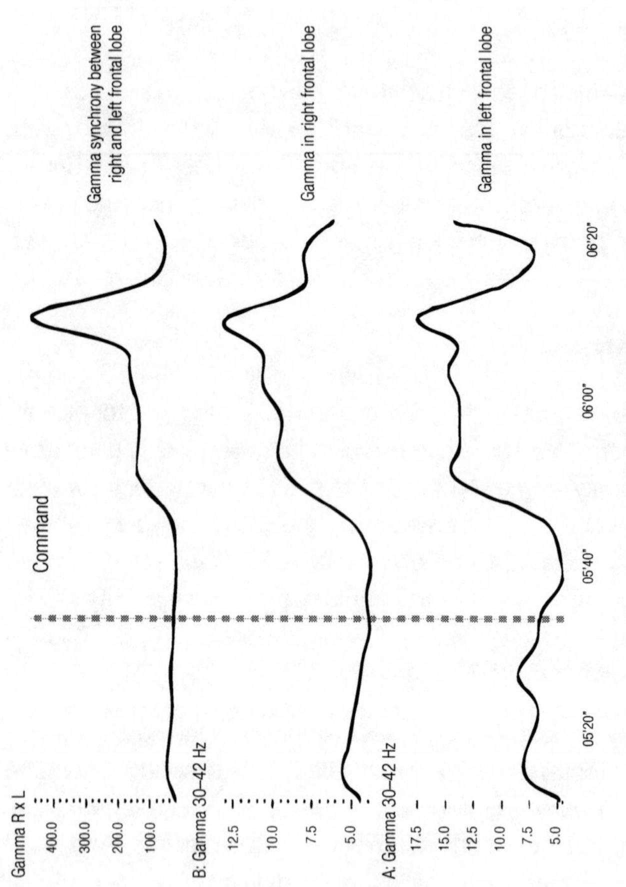

Figure 4.1: Subject Raises her Kundalini *Energy on Command*

at a maximum. This means that half a minute after the command was given, both frontal lobes are highly energized and totally synchronized. A high degree of synchronization (coherence) between the brain waves in the left and right brain reflects intensive communication or contact between the two cerebral hemispheres.

KUNDALINI AND THE FRONTAL LOBES

According to some authors, most of the frontal cortex is sleeping and it only functions mechanically, with little or no consciousness. In order to activate it, more energy is needed. Swami Satyananda writes: 'In order to arouse the silent areas, we must charge the frontal brain with sufficient *prana*, with sufficient vital energy and consciousness.' (Satyananda 1996: 23) He argues that this can only happen through a *kundalini* awakening.

If the frontal cortex were supplied with huge amounts of energy from the *kundalini*, it would probably lead to greater coordination and coherence of the neurons, thus improving the neurological basis for a higher consciousness. I suggest that if millions of neurons were to start synchronizing in unison at a high speed, very fast brain waves (gamma waves) would build up in the frontal lobes.

Interestingly, this hypothesis can be related to modern physics viewing the highly energized brain as a *Bose-Einstein condensate*. A condensate is a substance whose molecules have a greater order and coherence than normal. In a Bose-Einstein condensate the molecules not only have a high degree of order but also overlap, and the condensate behaves like a unit. Examples of Bose-Einstein condensates in inorganic matter are superconductors (conductors with no electrical resistance) and laser beams (beams of coherent light). A physicist by the name of Herbert Fröhlich (1983) has also shown that feeding extra energy to the cell walls of living tissue results in the cells lining up in the most orderly condensed way we know – a *Bose-Einstein condensate* – and the cell walls start acting as a unit.

In order to understand the power of coherence and synchrony, we shall turn to an old metaphor. If a regiment of soldiers walks at random across a bridge, nothing will usually happen to the bridge. However, if they march in step, there is a risk that the bridge will crash. Such is the power of coherence. Similarly, a laser light, where all the photons are completely aligned and the light is coherent, has such power that it can burn through thick iron plates.

Imagine the enormous brain power that would emerge if billions of brain cells were synchronized and totally coherent in the frontal cortex – the brain would behave like a Bose-Einstein condensate, becoming into a 'superconductor' with a focus like a laser beam.

SYMPTOMS OF *KUNDALINI* AWAKENING

There are many reports describing different sensory phenomena occurring during a *kundalini* awakening, often in the initial stages. Up to a few years beforehand, stirring sensations may be felt at the base of the spine. There can also be tickling sensations, often starting in the toes, feet and legs (predominantly on the left side) and then moving upwards toward the head. Hot and cold sensations may occur in different parts of the body. Commonly a variety of inner noises may be heard, such as whistling, hissing, chirping and roaring sounds. These internal sounds were reported by most *kundalini* subjects studied by Lee Sanella (Sanella 1992: 96).

Itzhak Bentov has the following description of the classic *kundalini* symptoms:

> *The sequence of bodily symptoms usually starts at the left foot or toes, either as a mild tingling stimulus or as cramps. The stimulus continues up the left leg to the hip. In extreme cases, there is a paralysis of the foot and of the whole leg. Loss of sensation in large areas of the skin of the leg may occur. From the hip the stimulus moves up the spine to the head. Here sometimes severe headaches (pressure-like) may develop.*
>
> *(Bentov 1977: 174–5)*

The reason for these symptoms is that on its upward motion, the *kundalini* may encounter all kinds of impurities and blocks. These will eventually be cleared by its dynamic activity, but this can be a very painful process. However, Osho emphasizes that *kundalini* symptoms are only felt if the system is blocked or if the person is resisting the energy. American psychiatrist Lee Sanella has also suggested that most, if not all, of the problems evoked by the rising *kundalini* are caused by conscious or unconscious resistance to the process.

Gopi Krishna writes in his autobiography *Living with Kundalini* about the problems associated with his awakening. He experienced pain, anxiety, psychotic episodes and thought that he would die:

> *It was variable for many years, painful, obsessive, even phantasmic. I have passed through almost all the stages of different, mediumistic, psychotic, and other types of mind; for some time I was hovering between sanity and insanity… I have passed through these stages, and then slowly my condition stabilized.*
>
> *(Krishna 1993: 124)*

Apparently when the *kundalini* is aroused, it can stir up repressed feelings and traumas that have been stored in the unconscious, causing a number of adverse symptoms, even psychotic episodes in predisposed individuals. The risk of a psychotic episode is probably greatest in people with unresolved early pain and trauma. If the *kundalini* rises suddenly in such an individual, it can be like putting 220 volts through a 110-volt system: all the weaknesses of the system will surface and create emotional and mental turmoil, with information overload in the brain and nervous system. This may cause anxiety, depression and psychotic episodes with delusions and hallucinations.

It has been estimated that a great number of patients diagnosed with schizophrenia or manic-depressive illness are really undergoing complicated *kundalini* arousals. The uncertainty about what is going on with them and the labelling of their condition as mental illness surely worsens that condition. In most cases, these patients do not need medication, but support and advice about what is happening to them.

This is what Swami Satyananda has to say on this matter:

> *Some of the people in the West who are locked up as insane would be recognized in the East as having undergone higher spiritual experiences. Therefore, it is now up to science to determine some definite, concrete and reliable ways to differentiate between the broken, insane mind and the opening, enlightened mind.*
>
> *(Satyananda 1996: 64)*

Gene Kieffer, president of the International Kundalini Research Foundation, comments on the maltreatment of people with *kundalini* symptoms, who are misdiagnosed as psychiatric cases:

> *All too often, however, persons diagnosed as schizophrenic are doused with Thorazine, Stelazine, Lithium, or a host of other chemicals designed to slay the kundalini serpent in its tracks ... knowledge of the Serpent of Fire can work miracles in the wards of our mental hospitals – and in the lives of those suffering from mental illness.*
>
> *(Foreword, Irving 1995: xxvi)*

On the other hand it is also well known that severe symptoms and diseases can precede and provoke a *kundalini* awakening. In his recent book *Out of the*

Darkness, British teacher Steve Taylor has described a number of cases where long periods of depression have preceded an awakening experience. Among the best known are those of Eckhart Tolle and Byron Katie.

Gopi Krishna suggests that the *kundalini* is responsible for both genius and insanity:

> *I claim that the commonly known abnormal and paranormal states of mind, such as retardation, neurosis, or insanity on the one hand, and exceptional talent or paranormal gifts on the other, all proceed from the working of the evolutionary mechanism – kundalini – and that, with advanced knowledge of this mechanism, the problems resulting from its malfunctioning can be cured or obviated, and the precious attributes which it evokes can be cultivated at will.*
>
> *(Quoted Kieffer 1983)*

Many people believe that following a full *kundalini* awakening, where the energy goes all the way up to the crown chakra, a person becomes awakened or enlightened and rises above all mundane problems. Generally, I try to avoid the term 'enlightened'; instead I prefer to use the word 'awakened'. In my vocabulary, *awakening* refers to a process where the *kundalini* has begun to move in the body and the person is beginning to have deep spiritual insights. This is a very subtle process, which eventually results in the rise of the *kundalini* energy all the way up to the brain. But even after such a full awakening, a person is not enlightened.

According to Swami Satyananda, after the initial awakening, the *kundalini* returns to dormancy time and again, due to karmic blocks in the system. Both personal and collective karma in the navel chakra (*swadhisthana*) block the energy. Following a *kundalini* awakening and the opening of the navel chakra, a person may be flooded with unconscious forces which give rise to a variety of mental or physical problems. Satyananda writes:

> *When the [*kundalini*] explosion takes place and swadhisthana begins to erupt, the aspirant is often confused and disturbed by the activation of all this unconscious material. It is totally impossible to understand these impressions, which are attributed to a disturbed mental condition… When* kundalini *is residing in* swadhisthana *chakra, the last vestige of karma is being thrown out and all the negative* samskaras *express themselves and are expelled. At this time you may be angry, afraid or full of sexual fantasies and passion. You may also experience lethargy, indolence, depression and all kinds of* tamasic *[slothful] characteristics. The tendency to procrastinate is very strong and you just want to sleep and sleep. This stage of evolution is known as purgatory.*
>
> *(Satyananda 1996: 136)*

The real hard work, integrating the sub- or unconscious, begins *following* a *kundalini* awakening. When this integration is completed, dreaming stops and the person becomes aware during sleep. Then it may be appropriate to call them enlightened.

KUNDALINI AND SEX

There's a great similarity between *kundalini* energy and sexual energy. Sexual energy traditionally goes downward in the body and out through the sexual centre. However, according to Eastern masters, there is another possibility, namely that the energy goes upward.

We normally see sexual energy as the driving force for reproduction of the species, but it seems that nature has also designed it for evolutionary purposes. Gopi Krishna has continually emphasized this:

> *By the arousal of* kundalini *we mean the reversal of the reproductive system and its functioning more as an evolutionary than as a reproductive mechanism.*
>
> *(Krishna 1975: 112)*

Osho also talks about the reversal of sexual energy:

> *The increase in sexual power and the opening of the* kundalini *passage are simultaneous – not the same, but simultaneous. The increase in sexual power will be the thrust to open up the higher centers; so sexual power will increase. If you can be aware of it and not use it sexually – if you do not allow it to be released sexually – it will become so intense that the upward movement will begin.*
>
> *(Osho 1976: 207)*

Many years ago Sigmund Freud and Carl Jung proposed a theory of sublimation of the libido into artistic talent and genius. Since then, many people have believed that celibacy can increase our energy levels and promote artistic, intellectual and spiritual skills. This has not been scientifically documented, but the reason why many monks and yogis live in celibacy is in order to conserve sexual energy and accumulate it for the purpose of spiritual development. Many yogis, including Gopi Krishna, believe that excessive waste of sexual energy may cause serious trouble, including psychosis. Krishna writes:

> *A depleted store of reproductive energy in an individual can prove seriously detrimental when* kundalini *is suddenly aroused. In fact, one of the reasons why a spontaneous activation of the Serpent Power often ends in mental disorder is the fact that, in addition to a faulty genetic heritage or unhealthy organic structure of the body, the excess expenditure of the reproductive essences can cause ravages in the system, which make adaptation to the new activity of the brain impossible.*
>
> *(Krishna 1993: 23–4)*

For most people, the sex act is just a release of tension. As Osho puts it: 'It's like a good sneeze.' Since we live in such a stressful world, sex is often misused in order to release tension from inner conflict. If a man feels an overflow of sexual energy, he usually wants to release it as quickly as possible in order to relax. However, if he ejaculates, there will be a waste of energy. Osho says: 'The feeling of emptiness that is overtaking the whole Western mind is just because of sexual wastage.' (Osho 1976: 208) However, both Osho and Gopi Krishna warn against repressing natural sexual urges and do not recommend celibacy unless it comes naturally.

Few attempts have been made to measure brain changes during orgasmic sexual activity. However, US researchers Harvey Cohen, R. C. Rosen and L. Goldstein found increases in theta waves, especially in the right hemisphere, in subjects approaching orgasm during masturbation (Cohen, Rosen, Goldstein 1976). I myself found similar EEG changes in a 54-year-old woman who masturbated and experienced an intense orgasm involving her whole body and mind. In her case, the EEG started slowing as she reached orgasm (theta), but after a short while it accelerated and high-amplitude fast activity (beta/gamma) appeared all over the brain.

It was interesting that immediately following the peak of orgasm, nearly all brainwave activity seemed to shut down. Both slow and fast EEG activity almost seized up for about five seconds and then alpha activity returned to a normal level (*see www.newbrainnewworld.com*). This strong reduction in all brainwave activity could very well represent the point in time when the rising energy reverses and takes a downward course. However, it could also be seen as an automatic mechanism protecting the person from energetic and emotional overload.

I believe that such an EEG pattern – large slow waves followed by large fast waves – is the brain's signature of an orgasm involving the whole body and mind. One may suspect that somewhat similar EEG changes would take place during a *kundalini* awakening. However, this is up to future scientific studies to find out.

There is a method called Tantric sex where you can enjoy sex without wasting the energy. If the sexual act is practised with meditative awareness and without loss of the seminal fluids, there is no loss of energy and the act is *not* followed by fatigue and a feeling of melancholy. On the contrary, vital energy is multiplied and conserved. When both partners melt into each other, energy is exchanged between the masculine and the feminine and builds to higher levels.

Osho says: 'Sex has to be transformed – neither repressed nor indulged. And the only possible way to transform sex is to be sexual with deep meditative awareness.' (Osho 1976)

Yatra da Silveira Barbosa discussed sexual energy and *kundalini* in the interview that I conducted with her in Amsterdam in 1999:

Q: Do you think that ayahuasca can arouse the *kundalini* energy in man?

A: *Yes. That is the reason why I suggest that people do not have sexual orgasms at least three days before and after the rituals or during the workshop 'Quest for Inner Truth'.*

Q: It is said that this slumbering energy resides in the root chakra, *muladhara*, and when aroused, it ascends through the spinal column, passing through each of the chakras, which are connected with different areas of the brain. Do you think it's possible that ayahuasca affects the chakras, facilitating the flow of energy and the awakening of *kundalini*?

A: *It certainly does. Ayahuasca is an upward journey through the chakras, and it consequently opens and clears the path for the* kundalini *to rise.*

Q: Gopi Krishna and others say that when the *kundalini* awakens it may bestow the individual with unusual abilities such as telepathy, clairvoyance and the capacity to have out-of-body experiences. Have you ever experienced such phenomena?

A: *Yes, ever since I was a child. It is as if I become a mirror and reflect other people's energy and situations. It's as if I am a radar with an antenna covering 360 degrees, and any movement in terms of energy is available to me from any direction. If somebody is thinking about me or directing their energy toward me, I often feel it. I know who is calling on the phone before I pick it up and I sense the arrival of a person before they are actually there. I also sometimes know a person is going to die before it happens.*

Q: According to Gopi Krishna, *kundalini* is an evolutionary biological force channelling transformed sexual energy through a reversal of the reproductive functions. Thus sexual energy should be preserved and utilized for the transformation of consciousness and for spiritual development. Do you agree with that?

A: *I agree with that when it happens spontaneously and with awareness, otherwise it becomes repression, and through repression there is no spiritual growth.*

Q: How does ayahuasca influence sexual functions and sexual energy? Is it possible to have an orgasm when you are high on ayahuasca? What would happen if you had one?

A: *Everything is possible. But if you had an orgasm in such a situation, you would just drain away all the energy that you needed for the transformation and there would be nothing left with which to awaken the* kundalini.

Q: Is it possible to have a *kundalini* awakening and a normal sex life at the same time?

A: *It depends on what you call 'a normal sex life'. You can practise Tantra, for example, and benefit from it in terms of transforming the sexual energy.*

Q: In order to awaken the *kundalini*, yoga, meditation, *pranayama*, celibacy, fasting, etc. have all been suggested. What method do you suggest?

A: *The one which is most harmonious to you. Many roads lead to Rome.*

THE SUBTLE BODIES: BODIES WITHIN BODIES

As we have seen, *kundalini* is said to flow through the etheric or energy body rather than the physical one. For more than a thousand years, Eastern yogic masters have claimed that in addition to the physical body, we also possess several of these so-called subtle bodies. Indian-born US physicist Amit Goswami says in his book *Physics of the Soul* (2001) that, besides the physical body, we possess other, more subtle bodies, which are less quantifiable and controllable.

What is the nature of these subtle bodies and where are they located? It seems that they are non-physical but not immaterial. They have vibratory frequencies well above the physical range, and that is why they cannot be measured and verified by any known scientific method.

An analysis of the physical body shows that it is made up of billions of cells, which in turn consist of molecules, atoms and electrons. In the final analysis we see that the physical body is not made up of dense matter but of vibrating electrically charged electrons and other elementary particles. In short, the physical body is made up of electrical vibrations and energy.

According to Osho, the subtle bodies lie at the atomic level in the physical body. To summarize the theory he proposes: as you go deep into the physical body you meet the etheric body, the home of the *kundalini* energies. If you go even deeper from there into the subatomic plane, you meet another subtle body, called by some the astral body, the home of thought (Osho 1984).

Some people, for example Amit Goswami, say there are five subtle bodies, but most yogis say there are seven. The principle is that the deeper you go into the microcosm of the physical body, the more subtle the bodies become, and the higher their vibratory frequency.

We must assume that the subtle bodies interact with the physical body and with each other, but we know nothing about the nature of this interaction. By postulating that the subtle bodies are continuous transitions between the physical body and our deepest level of being (the Atman or soul), we avoid Cartesian dualism. According to Osho (ibid), 'When the quantum physicists went from the world of matter to the sub-atomic world, they went – without knowing it – from the physical to the etheric plane.'

The implications of the subtle bodies are many. First and foremost it is important to emphasize that the physical brain is not the sole carrier of human consciousness. The subtle bodies can also carry Consciousness to the extent they are developed. Dream sleep and out-of-body experiences are examples of the subtle bodies being carriers of Consciousness while the physical brain is asleep.

The Etheric (Energy) Body

As the etheric or energy body (sometimes also called the emotional or vital body) is considered to be the subtle body closest to the physical body, and feelings are organized in the brain at a limbic level, the energy body is probably based on deeper brain structures. Thus, in order to access the energy body consciously, you have to break through any blocked feelings at a limbic level. This is what happens in primal therapy. Patients at the Primal Center who succeeded in releasing some of their emotional blocks felt much more spontaneous and alive afterwards and this may simply be because they got in touch with their energy body.

My research at the Primal Center clearly showed that when a patient succeeded in breaking through their emotional blocks, there was always a substantial increase in slow brainwave (alpha and theta) activity in their EEG. I believe that alpha waves reflect the ability to communicate with feelings and theta waves indicate access to the deeper levels of being – the subtle bodies. People who have no conscious access to their feelings, because of blocks, cannot access their energy body either. And that is also why they are more prone to disease.

When you meditate, you must take your awareness deep into the physical body and feel the subtle energy vibrations there. When you connect to this inner energy field, the cells of the body become more alive, the energy body is activated and you become more vibrant. The more attention you give to the energy field, the more intense and vibrant it gets. With a little practice you can easily feel that for yourself. Rub the palms of your hands against each other for a few seconds and then feel the subtle vibrations in the hands. As you focus on these vibrations, relax and breathe deeply, and the energy may spread to larger areas of your body.

When you take your attention deep into the body, you connect to the cellular level, the cells become more alive, and the energy body is activated. Your attention or awareness represents a sort of 'sun' in the universe that is your body. All the cells light up in the presence of this 'sun'. According to Eckhart Tolle, the energy body is the doorway to *life* and to who you are, just as it is the doorway to the subtle bodies. So by going deep into the body you can actually transcend the physical body and connect to the subtle bodies. This, I believe, is the primary aim of meditation.

Following a *kundalini* awakening, you become much more aware and conscious of your energy body. It is a type of blueprint of the physical body. If the physical body is damaged by accidents or disease, it is the intelligent action of the energy body that handles the repair and recovery of the physical damage. I also believe that the energy body is closely related to (if not identical with) the body's immune system.

Practising meditation or body awareness enhances the energy body and the immune system. It makes the body more resistant to physical illness and strengthens its ability to heal itself. Being consciously connected to the energy body is what makes you feel alive. My work tells me that the connection to the energy body is reflected in the EEG by high alpha and theta wave activity. In order to stay healthy, it is important that you always stay connected to this inner energy field, even when you are busy in the physical world.

A relaxed focus on the inner body strengthens and expands the energy body, while tension, stress and the suppression of feelings block the connection. When you feel fear, the energy body shrinks, and when you feel love, it expands. Every time you resist what is, it contracts, and you cut yourself off from your own source and from *life*.

My research indicates that the energy body is primarily accessed through the right hemisphere and the limbic system. I have shown that meditation, feeling release therapy and the intake of ayahuasca all primarily affect the right hemisphere. All three methods also stimulate the energy body. Following these activities, people become much more alive, spontaneous and present.

As you go deep in the physical body to the etheric body, the impact of the conscious mind increases. Negative thoughts may only be damaging to the physical body in the long run, but they can have an adverse effect on the etheric body immediately.

I believe that the adverse effects of stress and negative feelings can to some extent be contained within the etheric body, without 'spilling over' into the physical body. Therefore it is important to be conscious of your etheric body, as it can act as a buffer against stress. I have reasons to believe that such an etheric buffer system is reflected in a person's brainwave pattern in the form of increased alpha–theta waves.

Visualization can have a powerful influence on the etheric and other subtle bodies, especially if you are consciously connected to your deep levels. Actually, visualization *per se* is a way of connecting to the subtle bodies. When you visualize something, it takes on form in the subtle bodies, and if you hold the form long enough with sufficient intensity, it will materialize on the physical plane. This, of course, is the case with both positive and negative visualization.

As you take your awareness deep into the body (as is the case in meditation) and become aware of deeper and deeper levels, supra-normal powers and abilities may become available. Phenomena such as telepathy, telekinesis, clairvoyance and out-of-body experience may be possible (Osho ibid).

A CASE OF *KUNDALINI* AWAKENING

I was convinced that *kundalini* was real when a *kundalini* awakening took place in front of my own eyes in my partner Inger in India in January 2006.

Inger had previously suffered from severe depression because of the loss of a son many years before. She had tried medication and feeling release therapy and had engaged in intensive brainwave training for a couple of years prior to our trip to India. She had become adept at sustaining a very high alpha wave activity for more than an hour and her frontal gamma wave activity had also become exceptionally high through gamma wave training (*see p.142*). Using these techniques she was able to enter very deep states of meditation in which she experienced inner peace and harmony. Her baseline brainwave pattern looked quite normal except for an imbalance of alpha activity between the left and the right sides of the brain. Her alpha waves were twice as high in the left hemisphere as the right, which is typical of a person suffering from depression. However, following the *kundalini* awakening, her alpha waves were almost equally distributed between the right and the left hemispheres (*see www.newbrainnewworld.com*).

When we arrived in India, Inger was feeling pretty good and in harmony with herself. She had no expectations about what was going to happen to her there.

On the second day after our arrival we went by bus to an ashram in Nemam, north of Chennai, to see an awakened master called Bhagavan Kalki. Inger describes her experience in Nemam:

> *Thousands and thousands of Indians were there to see Bhagavan, so the place was pretty crowded. At one point our group of a few hundred people was stuck together in an enclosure in the meditation hall before getting in line to see Bhagavan. I saw a little Indian boy about seven years old crawling towards me begging for some rupees. I got pretty upset and told him with my eyes, 'No, don't beg here. Go away.' Immediately this anger was substituted*

by indescribable sensations of Grace and energy shooting up through my whole body and out of my head and eyes. *I now felt an overwhelming feeling of love toward this little boy, and I could not take my eyes off him.*

The explosion itself only lasted for a few moments, but it had tremendous implications for me. Right after, I could barely stand on my feet, but I made it to see Bhagavan, who was sitting in an armchair meditating, surrounded by five male and five female Cosmic Beings (his disciples). Everything in the room was totally illuminated and there was a radiance of Glory, which I don't have the vocabulary to describe. I was in total awe. I have never ever experienced anything that might give you just a hint of this experience. I was in awe. Tears were pouring out of my eyes.

'This was heaven!' I cried out on my way out.

I stumbled out in the open air again, crying, and somehow I managed to get myself onto the bus taking us back to our campus. On our way home I sat looking out of the window. Everything stood still. I realized how perfect the outer world was and the sensation of happiness and bliss was endless.

In the bus on the way home I was silent, just observing. Everything – I mean everything – was exactly as it should be. This was my experience, this was the truth.

In the evening, when I went to see my dasa *(Indian spiritual guide), my mind was asking questions:*

> *'What happened? What was all this?'*
>
> *My wonderful* dasa *looked at me with the most loving eyes I have ever seen. She said, 'You have been given the greatest gift of all. In India we call it a* kundalini *awakening.'*

Two days later in Golden City, approximately 200 women were gathered in the meditation hall in the morning. Everybody was to invoke 'presence' by meditating on God, Jesus, Buddha or the present moment, whatever suited them best. Here is what happened to Inger:

> *I was so filled up with thankfulness and the feeling of Grace that this automatically became my Presence. I was ecstatic with joy so I started dancing to the music, or rather my body started dancing. Other people went through deep feeling release and cried a lot, while others laughed and laughed. The energies in the meditation hall were very, very high.*
>
> *When the music stopped, we were asked to sit or lie down and meditate on Presence. All of a sudden I was whirled into a life review. I re-experienced all the negative emotions I had felt, especially toward my mother, but also in other relationships. Everything was turned upside-down, and I was now shown all the gifts I had been given: my mother giving birth to me and nurturing me, her love for me, my son giving his life to me so that I could experience how life-threatening pain could be turned into wisdom and love. Whatever had happened in my life was a gift and nothing was a coincidence.*

> *I cannot express the gratitude I felt. It was overwhelming and I cried and cried. 'How can I ever thank you?' was the thought that kept coming into my mind.*

The way Inger describes it, her life review seems to have been an intense self-therapeutic process where old memories and even traumatic events were viewed from a different angle. It washed away many deep-rooted belief systems, leaving her with a completely new view of reality. She told me afterwards she felt as if she was born anew and her life had completely changed.

I believe that such a radical change of view of one's past and one's reality can only take place on the basis of a simultaneous change of brain function. I suggest that Inger experienced a major breakthrough from her left to her right hemisphere, which was now able to give her its version of reality. This hypothesis is supported by the EEG measurements taken before and after the process in India: for the first time in many years, alpha activity in Inger's right hemisphere increased so much that it surpassed that of the left hemisphere.

After her *kundalini* awakening and life review, there was a newness, an originality and a lucidity to everything Inger experienced, and it was often accompanied by a child-like excitement, for instance when she was playing with animals or riding on a bus. Her inner life became much richer, especially as she felt strongly connected to her inner energies, which she felt throughout her body as a sweet, blissful stream of 'honey'. She actually said to me, 'If I ever lose touch with these energies again, I will lose myself – I will lose my life. Being connected to this eternal source, I feel at home, no matter what the outer circumstances.'

After these events Inger also seemed to be able to pass on these energies to others via *deeksha* (*see p.50*). She also exhibited a number of symptoms such as weakness, dizziness, headaches and pain in her body deep down at the cellular level. These were most pronounced immediately after the awakening.

She explained further when I interviewed her about her experiences in India:

Q: **What effect did your life review have on you?**

A: *I knew from then on that nothing was coincidental, that my life had been prepared for me by a higher intelligence behind all creation. All my belief systems were washed away and replaced by a sense of knowing.*

Q: **What do you mean by 'knowing'?**

A: *A feeling of absolute certainty – an experience and intuitive insight that cannot be questioned.*

Q: **Who is it that knows? Who is the knower?**

A: *Well, it is* not *the mind. Everything the mind knows is based on conditioning, learning and belief systems. The mind is full of knowledge, but it has no intuition. Instead, the knowing belongs to the real self, which is the intuitive part of us directly connected to God.*

Q: **What is it that you know for sure?**

A: *That there exists a source of all creation, which you may call God, All That Is, Cosmic Consciousness, the Ultimate Intelligence or whatever suits you. It has been and always will be, and we are all connected to it. Due to this fact the human species is one big family and you should treat all human beings as you want to be treated yourself. Therefore, erase the concept of enemy from your mind. The enemy you think is 'out there' is only a mind projection.*

Q: Have you become religious?

A: *No, not in the sense that there is someone or something outside myself that I worship, as is the case in traditional religions. I use the word 'God' simply because I cannot find a better one. But my God is not holy, rather he is part of me and he is my guide at all times. In order to become sane we need to reconnect to this source, otherwise we must face the possibility that our minds will destroy our planet and us.*

Q: When did you first feel that you were able to pass on energy to others – to give *deeksha*?

A: *Actually, right after the* kundalini *experience. It came as a flash of insight. Before, I would deny that anything like that could happen. I had absolutely no longing to do it. It was not my business.*

Q: What does it feel like to give or receive *deeksha*?

A: *The experience is more or less the same whether I give or receive it: I become almost intoxicated with energy and sometimes very dizzy from the sensations of blissful energies filling up my whole being.*

Q: How did you feel in the days and weeks after your awakening? Were there any adverse symptoms?

A: *It was a mixture – sometimes I was in heaven, sometimes my body and head were in extreme pain and I could not get out of bed. However, the doctor in Golden City had told me that this was part of the process and that the pain was due to the cells opening up and releasing old memories. I settled for that explanation.*

SAMADHI: THE RETURN TO THE CREATOR

When *kundalini* rises up the spine, it eventually reaches the brain, where it causes a state of rapture that sometimes turns into a full-blown experience of *samadhi* – the ultimate state of consciousness.

Such an experience has been described by Indian master Swami Satyananda who, after his own *kundalini* awakening, wrote:

> *When Kundalini Shakti reaches* sahasrara *(the crown chakra), that is known as union between Shiva and Shakti, as sahasrara is said to be the abode of higher consciousness or Shiva. Union between Shiva and Shakti marks the beginning of a great experience. When this union takes place, the moment of self-realization or* samadhi *begins… It is death of the mundane awareness or individual awareness. It is death of the experience of name and form… The experience, the experienced and the experiencer are one and the same … there is no multiple or dual awareness. There is only single awareness.*
>
> *(Satyananda 1996: 175)*

Osho describes three different states of consciousness: 1) Consciousness with a content (the normal mind state), 2) Consciousness without a content (the meditative state), and:

> *The third state is called* samadhi *– no content, no consciousness. But remember, this no-content, no-consciousness is not a state of unconsciousness. It is a state of super-consciousness.*
>
> *(Osho 2001: 54)*

Eckhart Tolle has more recently described his own transcendental experiences. In a state of depression, he thought, 'I cannot live with myself any longer' and, 'Is there one or two of me, the I and the self?' He was so stunned over this thought that:

> *...my mind stopped. I was fully conscious but there were no more thoughts. Then I felt drawn into what seemed like a vortex of energy. It was a slow movement at first and then it accelerated. I was gripped by an intense fear, and my body started to shake. I heard the words 'resist nothing' as if spoken inside my chest. I could feel myself being sucked into a void. I felt as if the void was inside myself rather than outside. Suddenly there was no more fear, and I let myself fall into that void. I have no recollection of what happened after that.*
>
> *(Tolle 1999: 1–2)*

Tolle's experience has aspects of both a *kundalini* awakening and a *samadhi* experience, even though it is not a description of a classic *samadhi* experience. Afterwards, as he explained:

> *I got up and walked around the room. I recognized the room, and yet I knew that I had never truly seen it before. Everything was fresh and pristine, as if it had just come into existence.*
>
> *(Ibid: 2)*

It sounds to me as if there was a change in Tolle's brain function, which shifted his consciousness from the left to the right hemisphere. That would explain why his mind stopped (a shutdown of the left hemisphere) and also why everything looked fresh and pristine, as if seeing it for the first time (activation of the right hemisphere).

I have already described my partner Inger's *kundalini* awakening in India. Six months after our return to Copenhagen, she attended a five-day conference in Stockholm, Sweden, with the Indian master Ananda Giri (one of Bhagavan Kalki's closest devotees), who was on a world tour with one of the Cosmic Beings from Golden City in India. She described what happened to her as follows:

> *On the third day everybody was to have a* deeksha *from the Cosmic Being. A couple of hundred people lined up for this event. When it was my turn I knelt down in front of the Cosmic Being, who put his hands on my head. I immediately experienced a sort of knock-out and felt I was passing out and falling.*
>
> *About half an hour later (I could tell from the number of people who had had* deeksha*), I returned to normal*

consciousness and found myself lying on the floor in the room. I tried to get up, but that was impossible. My body was dead, so to speak, while I was wide awake. I realized that something extraordinary had happened. While my body and mind had been gone, I had experienced an endless void, total emptiness or nothingness. Those were the qualities of the experience, and I was absolutely startled. What on earth was that? There was no subject, no object, in fact nothing*. Yet, I immediately understood that at the same time there was* everything.

Back home in Denmark, I wrote to my dasa *from Golden City and asked for an explanation. She wrote back that this was the experience of Pure Consciousness.*

When I interviewed Inger she explained more about how this experience had affected her:

Q: **Can you explain more precisely what *nirvikalpa samadhi* is?**

A: *Nirvikalpa Samadhi is experienced as a state of oneness beyond all form, time and space. One recognizes the Void, the absolute reality that everything is nothing. And you only know this in retrospect when after the experience you return to normal consciousness. In reality there is no one to recognize anything. You are included in the Void. You are yourself Pure Consciousness, or God if you like.*

Q: **Were there any significant life changes after your *samadhi* experience?**

A: *Yes, absolutely. The* kundalini *energies that had been working on my body and head 24 hours a day, sometimes to an unbearably painful extent, calmed down and in a matter of a few weeks they completely changed character. Since then, there has been an ongoing sensation of extreme bliss and happiness. It is as if my body and head are filled with a fluid as sweet as nectar, and almost every day I am compulsively drawn to watch it. The experience is beyond any description that I know of. This energy of sweet nectar is life itself.*

Q: What is your view now on the issue of free will?

A: *With the earlier-mentioned life review in mind, it is very clear to me that in the ordinary state of consciousness there is no such thing as free will. All our actions are based on old conditioning and belief systems. This, however, does not free us from responsibility, since there are lessons to learn. In the awakened state I believe we do have a free will, but if we are in a higher state of consciousness we will always act in accordance with our true needs, which are always aligned with the divine principles of love and compassion. Thus in the awakened state we always act in accordance with God's will. God's will is now our will.*

Q: What is your view on the issue of karma?

A: *Previous to my experience in India I had read a little about karma without really understanding what it was all about. Now I see clearly that our acts in our present state of consciousness have to be rewarded or the contrary, since our life on Earth is an educational period guiding us toward higher states of consciousness. So there is actually no punishment taking place, but guidance for our own good. Since we create karma with our ego, this creation will stop when we drop the ego and reach the awakened state, because then we will, as I said just before, be in accordance with divine principles. Ultimately, since we are all united on this planet, we are not only responsible for ourselves but also for each other, and this is why humanity also has a collective karma from which none of us can escape.*

Q: How is life after awakening? What has happened to your perceptions, memory, sense of time, etc?

A: *The mind is very quiet, there is no interpretation of the outer world and there are no comments on what I experience. Everything is perfect at all times simply because it cannot be otherwise as long as our present state of consciousness prevails. However, experiencing a higher level of consciousness is a tremendous relief, almost like coming out into the sun after having been held prisoner. As to my memory of the past, it has totally lost its importance. Most of the time I live in the present moment. Regarding the future, my mind is very fast and effective when it comes to planning a journey, for example, but not when planning the content of a workshop. I just show up and then everything happens intuitively.*

Q: Do you still have feelings? For example, can you get angry?

A: *Yes, I can get angry, but the feeling is short-lived. I hardly feel the energy before it is gone again. It is like a wave rising and falling down again and there are never any residues. I also feel other people's pain. However, there is no identification with my feelings or with my thoughts. I just watch and accept whatever is there.*

Q: Do you still have an ego? Has your former personality changed?

A: *I have no idea whether I have an ego or not. I think that my personality from the outside looks pretty much the same as before. However, the urge to be of help to other people has become my number one priority. I have therefore engaged in humanitarian work as a volunteer for the Danish Refugee Organisation and the Rehabilitation Centre for Victims of Torture.*

Q: In your opinion, is there anything one can do to awaken?

A: *I am not the right person to answer this question. I feel I did not prepare for the* kundalini *awakening. It hit me and I have no explanation why. Some people have suggested that the feeling release therapy I did, especially the intensive gamma brainwave training I did over the last few years, prepared me for the experience in India, but I don't know about that. I think, however, that if you want to grow spiritually, you will benefit a lot from doing selfless service. That will help you much more than studying books about spirituality. Books provide you with knowledge that may be useful to some extent, but you will learn much more by experiencing for yourself.*

Q: Being in the awakened state, are you a perfect person living a perfect life?

A: *What does it mean to be perfect? Babies and small children gifted with parents who love them unconditionally are perfect individuals living a perfect life – joyful, playful, innocent, trusting, etc. My life is in many ways very childlike and I experience life, love, joy and peace at all times. Occasionally I have health issues, but they do not concern me very much. The body is temporary, as is everything except life itself, but the body is also the most precious gift we have ever received and we should treat it accordingly.*

Q: How would the world change if most people were in the awakened state?

A: *If you function at a higher level of consciousness it is* not *possible to fight, nor is it possible to exploit other people. Your overall concern will be the well-being of other people and of the planet. Try to imagine for yourself what the implications would be for humanity and for the Earth if most people were in such a state.*

Individuals have been experiencing similar states throughout history. In an article co-authored by Charles Tart, 38-year-old Californian resident Allan Smith described a *samadhi*-like experience, which he called an experience of 'Cosmic Consciousness', that he had one evening in 1976:

> *There was no separation between myself and the rest of the universe. In fact, to say that there was a universe, a self, or any 'thing' would be misleading – it would be an equally correct description that there was 'nothing' as to say that there was 'everything'. To say that subject merged with object might be almost adequate as a description of the entrance into Cosmic Consciousness, but during Cosmic Consciousness there was neither 'subject' nor 'object'. All words or discursive thinking had stopped and there was no sense of an 'observer' to comment or to categorize what was 'happening'. In fact, there were no discrete events to 'happen' – just a timeless, unitary state of being… Perhaps the most significant element of Cosmic Consciousness was the absolute knowingness that it involves. This knowingness is a deep understanding that occurs without words. It was certain that the universe was one whole and that it was benign and loving at its ground.*
>
> *(Smith, Tart 1998: 100–101)*

Osho believes that the state of *samadhi* involves an opening up of the bridge between the hemispheres and a merging of the left and right hemispheres:

> *If this bridge is strengthened so much that the two minds disappear as two and become one, then*

> *integration, then crystallization, arises. What [the Russian philosopher] George Gurdjieff used to call crystallization of being is nothing but these two minds becoming one, the meeting of the male and the female within.*
>
> *(Osho 2001: 127)*

Thus, Osho says, during *samadhi* there is a union between Shiva and Shakti, the male and the female principle, which is reflected in the merging of the left and right hemispheres.

Following this analysis of *samadhi*, it seems obvious that it is *creation in reverse*: the ego is lost, subject and object disappear and there is a feeling of oneness with God.

Osho also states that it marks the return to the creator:

> *[Samadhi] is a state of super-consciousness, of transcendental consciousness. Consciousness now is only conscious of itself. Consciousness has turned upon itself; the circle is complete. You have come home [my emphasis].*
>
> *(Ibid. 54)*

COSMIC CONSCIOUSNESS

Canadian-born psychiatrist Richard Maurice Bucke coined the term 'Cosmic Consciousness' in his 1901 book of the same name. According to Bucke, there are three major levels of consciousness in evolution. First, there is Simple Consciousness, which is possessed by the upper half of the animal kingdom; next there is Self-Consciousness, which is possessed by most humans; and finally there is Cosmic Consciousness, the highest level of consciousness, which is only prevalent among a very few humans on the planet.

Bucke himself had an experience of Cosmic Consciousness when he was 36 years old. He and two friends had spent the evening reading Wordsworth, Shelley, Keats, Browning and Whitman. When driving home from the meeting, Bucke (referring to himself in the third person) found:

> *He was in a state of quiet, almost passive enjoyment. All at once, without warning of any kind, he found himself wrapped around as it were by a flame colored cloud. For an instant he thought of fire, some sudden conflagration in the great city; the next, he knew that the light was within himself. Directly afterwards came upon him a sense of exultation, of immense joyousness accompanied or immediately followed by an intellectual illumination quite impossible to describe. Into his brain streamed one momentary lightning-flash of the Brahmic Splendor, which has ever since lightened his life; upon his heart fell one drop of Brahmic Bliss, leaving thence-forward for always an aftertaste of heaven.*
>
> *(Bucke 1901: 10)*

This description is akin to other descriptions of *kundalini* awakening (cf. Gopi Krishna), however, it has little similarity with the above descriptions of *samadhi*. I suggest that what Bucke calls Cosmic Consciousness' is an awakened state following *kundalini* arousal.

In his book Bucke collected an impressive sample of

cases of people who had supposedly achieved Cosmic Consciousness, including Buddha, Jesus, Paul, Dante and Whitman. These are the characteristics he believes are associated with it:

> *The prime characteristic of Cosmic Consciousness is, as its name implies, a consciousness of the cosmos, that is, of the life and order of the universe... Along with the consciousness of the cosmos there occurs an intellectual enlightenment or illumination which alone would place the individual on a new plane of existence – would make him almost a member of a new species. To this is added a state of moral exaltation, an indescribable feeling of elevation, elation and joyousness, and a quickening of the moral sense, which is fully as striking, and more important both to the individual and to the race than is the enhanced intellectual power. With these come, what may be called a sense of immortality, a consciousness of eternal life, not a conviction that he shall have this, but the consciousness that he has it already.*
>
> *(Ibid: 3)*

It has been claimed that LSD and other psychedelic substances can induce ecstatic states of consciousness and even to varying degrees fulfil Bucke's criteria for Cosmic Consciousness (Stolaroff 1979). Following Allan Smith's spontaneous experience of *samadhi* or Cosmic Consciousness, he was curious as to whether he could reproduce it by taking LSD. (Prior to this experience he had never taken LSD, but had occasionally smoked marijuana.)

He did manage to induce a transcendental experience with LSD, but:

> *The mood changes were both quantitatively and qualitatively different. During LSD intoxication, my mood was brittle. A 'high' state could be rapidly converted to a depressed one and vice versa ... the mood swings were infrequent. In contrast, the Cosmic Consciousness mood elevation was constant, solid and all-pervasive. It was so intense that the words 'joy' and 'high' fail to capture the experience. The mood elevation of Cosmic Consciousness and a positive LSD peak not only differ in intensity, but have a different feel. I would describe the LSD experience as 'high' and Cosmic Consciousness as 'ecstatic'. The best verbal description that I can give for the difference between the two is the extent of associated ego loss ... the self was never far away with LSD, but was totally beyond recall with Cosmic Consciousness.*
>
> *(Smith and Tart 1998: 104–5)*

Thus, according to Allan Smith's description, the LSD-induced high was a much more unstable state with mood swings and only partial ego loss compared to the spontaneous Cosmic Consciousness, where the ego was lost completely.

Chapter 5

AWARENESS AS A DRIVING FORCE IN CONSCIOUSNESS DEVELOPMENT

Awareness and attention are key concepts in brain evolution. Gopi Krishna's view was that: 'Attention or concentration of mind is the instrument by which nature accelerates the process of evolution.' (Krishna 1975: 121)

Some researchers believe that attention is either identical to consciousness or causes it. I rather suggest that the brain's attention mechanism is an important tool that is necessary for human consciousness to work properly.

According to Wikipedia, 'attention' is 'the cognitive process of selectively concentrating on one aspect of the environment while ignoring other things'. Thus it usually refers to objects in the external world. If, on the other hand, one is attentive to internal objects, such as thoughts, sensations and images, it is called 'concentration'.

It is important to discriminate between attention/ concentration and awareness. *Attention* or *concentration* involve focusing on an outer or inner object or on some details to the exclusion of everything else (this is also called *selective attention*), while *awareness* means being open and sensitive to everything around you and inside you. Concentration is primarily a left-hemisphere function, while awareness is a right-hemisphere facility. Concentration also involves effort, while awareness is effortless.

The EEG actually discriminates between these two qualities of mind. Since concentration involves effort, it will suppress alpha activity, while awareness, which involves an open, passive but sensitive mind, tends to increase it (*see Figure 5.1*).

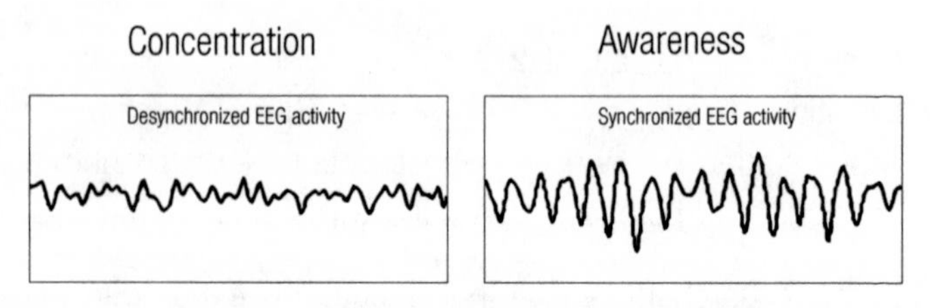

Figure 5.1: Concentration versus Awareness

This is what Osho has to say about concentration versus awareness:

> *...the mind must be able to concentrate. But the moment the mind becomes able to concentrate,*

> *it becomes less aware. Awareness means a mind that is conscious but not focused. Awareness is a consciousness of all that is happening. Concentration is a choice. It excludes all except its object of concentration; it is a narrowing… The moment you narrow the mind you become particularly conscious of one thing and simultaneously unconscious of so many other things. The more narrowed the mind is, the more successful it will be. You will become a specialist, you will become an expert, but the whole thing will consist of knowing more and more about less and less.*
>
> *(Osho 196: ix)*

He also emphasized that awareness was enough for consciousness development, saying, 'Evolution is between awareness and unawareness. Becoming more aware and less unaware.' (Osho 2005: 55) Just being 'the watcher on the hill' observing everything that goes on in the mind without identifying with it will take you to a higher level of consciousness.

Meditation is awareness training rather than concentration training. However, initially it can be very useful to practise sustained one-pointed concentration. Either attention is directed towards the body (for example to breathing) or towards the environment (for example to a spot on the wall). In the first case the eyes are usually closed and the EEG shows slow synchronized waves (alpha), while in the second case the eyes are open and the brain waves are synchronized at a higher frequency (beta or gamma). Following such concentration exercises it seems easier to open up your focus in all directions and become aware of the totality within and without.

The ability to sustain such an open focus of attention over prolonged periods seems to be the crucial factor in meditation. It is interesting to note that if you are concentrating, for example on a spot on the wall, initially your focus is very narrow. After a short while, however, the spot will appear blurred because your eyes will become accustomed to looking at it. At this point your focus will become more open and eventually concentration will turn into awareness – of yourself and/or of the environment.

It is important that as you concentrate on an object or a person, you are simultaneously aware of yourself. Russian philosopher Gurdieff often emphasized this, saying, 'Remember yourself.'

This double-arrowed consciousness, where you are simultaneously aware of the object and the subject, is actually the same as presence. Osho comments:

> *In yoga the effort is to become conscious of both the object and the source. The consciousness becomes doubled arrowed. You must be aware of the object, and you must be simultaneously aware of the subject. Consciousness must become a double arrowed bridge. The subject must not be lost, it must not become forgotten when you are focused on the object.*
>
> *(Osho 1976: 2)*

It is likely that intensive training of attention/awareness over long periods of time affects brain function. American neuro-psychiatrist Jeffrey Schwartz has argued that the mental effort exerted during attention training releases a 'mental force' that stimulates the brain's ability to create new synaptic connections. Using PET scans he has demonstrated that attention training increases blood flow to certain areas of the brain (Schwartz and Begley 2003: 7).

I have no doubt that systematic attention training has a powerful influence on brain function and it seems that it may even affect single neurons. Let me suggest a metaphor: I believe that concentrated attention increases the coherence of brain cells just as iron filings align themselves in the presence of a strong magnet. The more intense the concentration, the stronger the coherence between the brain cells.

In the busy, stressed and restless world of today, one-pointed concentration training does not seem to occur naturally. On the contrary, wherever we go, the mind seems to be constantly stimulated to change its focus. Almost everything and everybody we encounter in the civilized world wants our attention: newspaper adds, TV commercials, salesmen, etc. Computer action games, action movies and certain TV programmes that incessantly jump from one scene to another challenge our capacity to shift our attention rapidly to the utmost limits. This is sensory overload aimed at speeding up brain activity and providing an adrenaline kick, but at the same time it is almost raping our attention mechanism. We should warn our children and youngsters against indulging in such activities, which, according to some psychologists, may cause attention disorders such as ADHD. At a conference at MIT a cognitive scientist remarked, 'In current times, people are suffering from a disorder I call MAD ... I mean they are suffering from *Multitasking Addiction Disorder*.' (David Meyer, University of Michigan, quoted Harrington and Zajonc 2006: 52)

Furthermore, humans want attention from other humans. Both energy and money (also a kind of energy) follow attention. Look at rock and movie stars – they always get a lot of attention, and money as well. If you can attract attention, you can also attract money. Receiving attention from others empowers you, just as attention from yourself will empower the cells in your body and stimulate the energetic system (energy body). With the development of the internet, information has become an easily available resource for most people. What we need now is more attention and awareness in order to digest all the information.

Herbert Simon (1971) seemed to be the first person to articulate the concept of attention economics. He says that in an information-rich world, the huge amount of information leads to a lack of what information consumes, namely attention. Thus, a wealth of information creates a scarcity of attention and a need to direct that attention to the proper information.

Simon noted that many designers of information systems incorrectly represented their design problem as information scarcity rather than attention scarcity. As a result they built systems that excelled at providing more and more information to people, when what was really needed was a system that filtered out unimportant or irrelevant information. (Simon 1971: 143–4)

In *The Attention Economy*, Harvard psychologists Thomas Davenport and John Beck argue that knowledge and information are not scarce in our modern society and emphasize that what we are short of is attention, which is now the most resourceful commodity. They also claim that 'understanding and managing attention is now the single most important determinant of business success'. (Davenport and Beck 2001: 3)

What do you pay with when you pay attention? 'Energy' is the obvious answer to that question, but how?

'Neural mirroring' may explain how. Researchers in Italy found out that when a monkey ate a peanut in the laboratory a certain electrical brain pattern was produced. Interestingly, that very same pattern was reproduced when the monkey was just observing a human eating something (Rizzolatti, Craighero 2004). This is an example of 'neural mirroring'. Thus, when you are paying attention to somebody it looks as if you are reshaping your mind (and your brain) to mirror that person's state of mind. This mechanism, by the way, can also explain how you empathize with another person and recognize their intentions.

In order to pay full attention to another person you need to be present in the moment. If you have other agendas, thoughts or feelings, you are not totally present. Imagine that your whole brain (both hemispheres) and body are focused on a specific person at a specific moment – then you are totally present with your whole being.

The most outstanding feature of an awakened master, compared to an ordinary person, is no doubt their degree of presence. A person's presence can usually be felt. It has nothing to do with what they say or how they look, but there is a certain vibration or charisma around them, especially in their eyes, which cannot be explained. As Ram Dass says, 'The only thing you can really communicate to another person is your being.' (Dass 2004: 5)

The opposite of presence is, of course, absence. While the brain of the absentminded person is characterized by too many slow brain waves (theta waves), the brain of the intensely present person is characterized by a lot of fast, coherent brain waves (beta or gamma waves), especially in the frontal part of the brain. When you are present in the moment, a higher level of consciousness is in charge, but when you are absent, lower (unconscious) levels dominate.

BRAINWAVE TRAINING: A NEW CONSCIOUSNESS TECHNOLOGY

There is a relatively new and promising method of attention/awareness training called EEG biofeedback, or neurofeedback. This is actually technology-supported awareness training. In a sense, it is a Western scientific approach to meditation.

I have tried to emphasize how important it is to be able to direct, harness and control attention. This was also the point of the great American psychologist William James when he wrote, more than 100 years ago:

> *The faculty of voluntary bringing back a wandering attention, over and over again, is the very root of judgement, character and will... An education which*

> *should improve this faculty would be the education* par excellence. *But it is easier to define this ideal than to give practical instructions for bringing it about.*
>
> *(James 1890: 424)*

James emphasizes the importance of training attention and calls for a practical method of training it. I think that neurofeedback is such a method. It can effectively teach you to discipline your attention and thinking, which in turn will improve your meditation, among other things. We have had a number of clients at our centre in Copenhagen claiming that it was only after training with neurofeedback that they really understood what meditation was.

Neurofeedback training is based on a feedback signal to the subject reflecting the state of their brain. Small sensors placed on the scalp pick up their brain waves, which are then amplified and fed to a computer. A software program analyses the signals and converts them into sounds, music and animations, which are fed back to the subject via loudspeakers and a computer screen. The feedback signals help the subject become aware of even the slightest change in their brainwave pattern and thus also in their state of consciousness. If they become inattentive or drowsy or start thinking, their brain waves will change accordingly and they will know immediately because the feedback signal will tell them.

Many types of brain wave can be trained with neurofeedback methods. Alpha wave training increases body awareness and reduces stress, beta wave training increases concentration and gamma wave training improves focus, presence and stamina among other things. Poor concentration is often accompanied by too many slow theta waves in the EEG, while good concentration is accompanied by fast brain waves such as beta or gamma. Adults and children with attention disorders usually have too much slow activity and too little fast activity. It is a good idea to train these persons to inhibit their slow theta waves and increase their fast beta waves.

Of whatever type, neurofeedback training is primarily attention training. When our attention is directed towards the outer world, the brain normally cycles at fast beta wave frequencies, but when attention is directed towards the body, it slows down and produces alpha wave frequencies. Thus it makes sense that in order to increase body awareness you should train alpha waves, while in order to promote outward concentration you should train beta or gamma waves.

You may ask how neurofeedback training works. On the physiological level, the intensive training of attention seems to stimulate neurological coherence and promote the general maturation of the brain. In a sense you could say that neurofeedback teaches the neurons in the brain to co-operate. This occurs during training when the neurons are stimulated to create new synaptic connections, thus improving brain functions in the areas concerned.

There is evidence that both alpha and beta training increase blood flow to the brain. Nagata (1988) found a negative correlation between delta/theta activity and blood flow and a positive correlation between alpha activity and blood flow in the brain. In other words, when a person succeeds in going into a high alpha state, their brain receives more blood and therefore more oxygen and nutrition. This, of course, stimulates impaired neurons to regenerate and improve their conductivity for electrical signals.

'Neuroplasticity' is a term for neurons' ability to create new synaptic connections and pathways in the brain. A good example of this is when a new area in the brain takes over the damaged functions of another area. Presumably neuroplasticity also comes into play during neurofeedback training.

From a psychological point of view, neurofeedback training is a process that increases awareness of both body and mind. For example, during alpha training the person becomes more attentive to their body, feelings and impulses. Gradually they learn to confront and observe, passively, their tensions, restlessness, feelings and compulsive thoughts. If they give up their resistance to these inner impulses and observe them with full attention, they will usually fade away. This is a process that learning psychologists call 'deconditioning' or 'extinction'.

Training for Body Awareness and Grounding (Alpha Training)

Alpha waves reflect the brain's idle function in the awake condition – a conscious state of being without any doing. They represent a gate between the outer and the inner world – and between the conscious and the unconscious. When we relax with closed eyes, most of us produce a certain number of alpha waves. The moment we start doing something – concentrating, thinking, listening, etc. – the alpha waves are blocked or reduced in amplitude (fewer microvolts). Also, if we get drowsy, alpha activity drops. Thus, in order to stay in the high alpha state, we must be relaxed but at the same time very alert and attentive.

When we do alpha training, the subject sits in a chair with closed eyes and listens to a deep tone representing alpha activity. Whenever alpha waves are present in his EEG he will hear the tone, and if the alpha waves drop down, the tone will disappear. Thus, after some practice through trial and error, he may learn to stay for long periods of time in the high alpha state while the feedback tone is on almost continuously.

Alpha waves reflect a calm, open and balanced mind with a free flow of energy in the brain. Alpha activity also reflects conscious access to the body and its feelings, thus indicating good body awareness. Athletes and people in good physical condition tend to have alpha levels well above average. Physical and mental tensions usually block the energy and reduce alpha activity. Permanently blocked alpha activity may lead to various stress symptoms and burnout.

In order to produce alpha waves you must control your attention and direct it toward your inner body, ignoring all outside stimuli. When thoughts appear, you should not follow them or start co-operating with them. Ignore them or see them for what they are and let go of them, just as you let go of tension in your body.

During alpha training you consciously let go of your inner tensions, thereby dissolving blockages. It is as if tensions and blocks dissolve in the light of your consciousness. The energy follows your conscious attention and the process is accompanied by increasing alpha activity.

The most effective way to increase your alpha level is to give up all resistance and surrender to the present situation. However, it is important that this is done with full awareness; if you become drowsy, the alpha waves will drop instantly.

Alpha training, then, is conscious and systematic focusing of mental energy. It is a charging of the brain with new energy, which is then available for intellectual and creative activities. As I have already said, research indicates that the blood supply to the brain increases in the high alpha state and the immune system is also strengthened.

Alpha wave training is effective against stress because it attacks problems at their roots. It removes or reduces the habitual tendency to tense, resist and block the energy in response to stressful situations. While other methods of stress management usually try to establish new behaviour patterns, alpha wave training focuses on eliminating the inappropriate patterns and habits that led to the stress condition. While many methods try to *reprogram* you, alpha wave training will actually *deprogram* you.

Alpha wave training heightens consciousness and makes you more aware of yourself, your feelings and your body. It also teaches you to stay relaxed and focused in times of stress, and I believe it is the most effective way to learn to meditate properly. Since research has documented that deep meditation is usually associated with high alpha activity in the brain, the high alpha state has become almost synonymous with meditation. When a person is

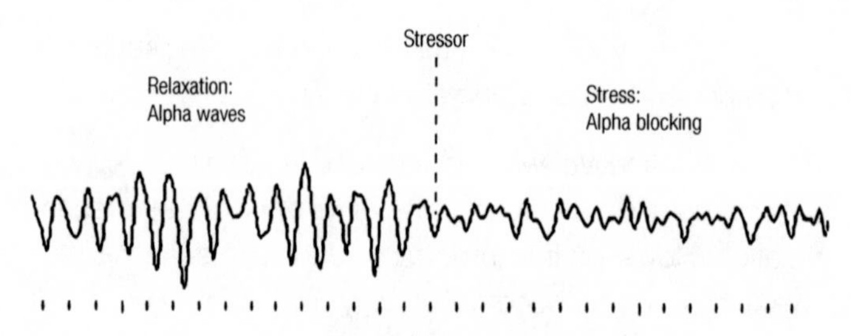

Figure 5.2: The Brain's Response to Stress

unable to meditate properly, stress is usually the problem. If the person is tense, has a busy mind or resistance toward many things, alpha activity is inhibited and it's impossible for the person to meditate.

A person with a high alpha level has an efficient brain and goes without effort from one task to another. It is interesting to notice that virtually all children up to the teenage years have lots of alpha waves. Many adults, however, suppress their alpha over the years, mostly due to the accumulation of stress.

I believe that people with high alpha levels have more resilience towards stress than people with no or low alpha activity. If you resist your stress impulses, consciously or unconsciously, your alpha waves will block and the tensions will remain in the body without an outlet. This is called repression and in the long run it can harm the body and lead to many diseases.

Usually the biggest problem when you do alpha wave training is controlling your thinking. The moment you start thinking – and that often happens unconsciously – the alpha waves diminish or disappear. It can be very difficult to stop thinking, especially if you are under stress. If you find this is the case, you should not try to repress the thoughts but instead observe them passively for a moment and then ignore them. The trick is not to let any thoughts sneak up on you because you are not fully alert and conscious.

If you are very skilled at alpha training, however, and can sustain a high alpha state for a long time, it is possible to start thinking without reducing your alpha waves. Then almost every single thought that goes through your mind will be conscious. That, of course, is the ultimate target. When you have reached that state of consciousness, you are using your thoughts; they are not using you anymore.

People with busy minds often have great difficulty in controlling their thoughts. In this case we initially let them train fast gamma waves from the frontal lobes. Frontal gamma training, explained in the next section, is very effective in clearing the mind of thoughts. After, say, 20 minutes of gamma training, your thinking mind will be put to rest and it will usually be much easier to do alpha training.

At our centre we have had very good results with the combination of gamma and alpha training. I believe that alpha wave training improves emotional intelligence while frontal gamma training facilitates focus and presence and also very likely stimulates creative and intuitive abilities.

Training for Focus and Presence (Gamma Training)

Gamma waves are actually very fast beta waves, with frequencies up to and around 40 cycles per second. The 40 Hz gamma activity was originally studied by Professor John Jeffreys at the Neuroscience Unit at the University of Birmingham. The scientists there found these frequencies to be associated with higher levels of brain organization, 'binding' information from all the senses together in one whole conscious experience.

Since gamma rhythms disappear under general anaesthesia, they seem to be associated exclusively with higher mental activity and consciousness. They can be seen all over the brain, but are most prominent in the prefrontal cortex, which is the seat of the executive functions such as the direction of attention and the regulation of emotions.

Gamma waves may be seen as the muscles of the brain. Whenever you really need to achieve the utmost, physically or mentally, through intense focus, gamma waves are called for.

From the work we have done at the Mental Fitness and Research Center in Copenhagen we have gathered evidence that gamma waves are associated with will, focus and unity of experience. In a number of subjects we have found that frontal gamma training is the most effective way to increase a person's ability to focus. Also, we have found indications that this training increases energy levels, willpower and the ability to stay present in the moment.

The monks studied by neuroscientist Richard Davidson (Lutz and Davidson 2004) had unusually high levels of gamma in the frontal area, especially when they meditated, but also at rest.

We ourselves have had top athletes in our training who showed very high gamma levels to begin with and were able to increase these further through neurofeedback training. One fine athlete who had won the world canoeing championship 10 times was able to increase the gamma activity in the frontal area by a factor of seven in 20 minutes. After the training session he told us that when his gamma was at a maximum he was in the same state as when he was in a canoe race close to and completely focused on the finishing line.

When we do frontal gamma training, we have a person sit in front of a computer screen watching a bar graph showing how their gamma level is rising or falling according to the intensity of their concentration.

After several training sessions we have the subject focus on a blue star on the wall with a high-pitched tone as the only means of feedback. If they stay intensely focused, the tone will be on most of the time. When they have observed the star on the wall for a few minutes, it will begin to appear blurred because their eyes will have adapted. At this point there is a tendency for the object of their focus to meld with the subject's forehead. In a sense the blue star dissolves into the person's mind and the person then becomes acutely aware of themselves.

It usually takes 10–15 minutes of training before people reach a preliminary maximum of gamma activity. In successfully trained people, we expect to see a steady increase of gamma over perhaps 20–30 minutes. However,

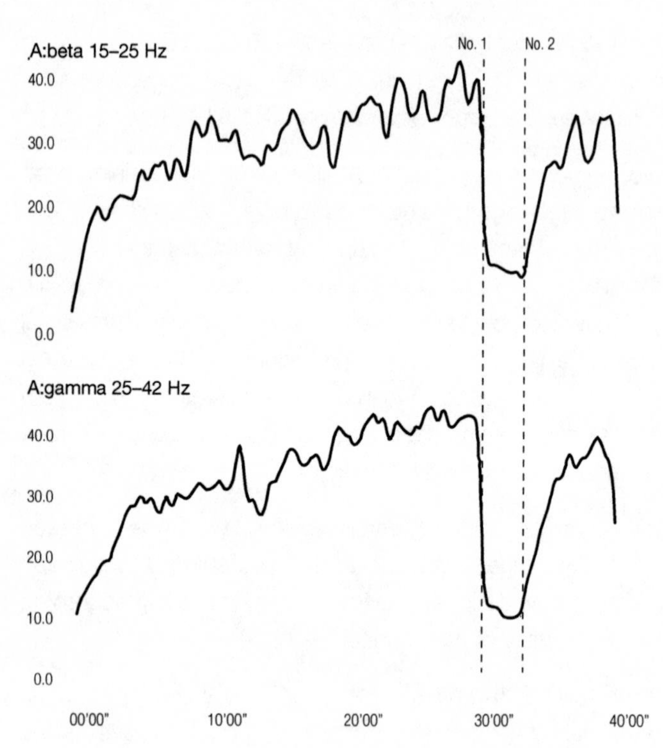

Figure 5.3: Unobserved (Unconscious) Thinking Shuts Down the Frontal Lobes

the moment the subject starts thinking, they will lose focus and the gamma level will fall abruptly (*see Figure 5.3*).

I have no doubts that gamma wave training activates the frontal cortex and raises the consciousness 'above' the thoughts. In this state there is no habitual, unconscious thinking. In fact, I believe that in the high gamma state it is impossible to think an unconscious thought. You can still decide to think, but all your thoughts will be conscious.

In the high gamma state you have a very lucid mind, and a penetrating focus, which sharpens your intuition and improves creativity.

After just 20 minutes of frontal gamma training most people experience a clear and quiet mind, sharp focus, increased energy and a relaxed body. Some even become euphoric.

It is interesting that in the old yoga treatises there are descriptions of how a state of unbroken fixity of attention can be achieved:

> *The target to be attained is that the observing mind and the object contemplated should fuse into one. This can occur in only two ways: Either the object dissolves into consciousness and only the seer remains intensely conscious of himself, or he loses his own identity and become one with the object on which the mind is fixed.*
>
> *(Krishna 1975: 44).*

This is exactly how most of our clients experience gamma training: after a while there is a fusion of the observing mind and the observed. This state is sometimes experienced as a light trance; however, the person is very conscious and alert and does respond immediately to questions. This is why I sometimes call the high gamma state a *conscious trance*. The word 'trance' only refers to the fact that awareness has been extremely narrowed.

I propose that in the high gamma state the thinking parts of the brain (parietal lobes) are shut down, while the frontal areas are activated. I also suggest that the unbroken high gamma state is a flow state in which there are no thoughts and no resistance. I also believe that it is closely related to both creativity and intuition.

Based on many years of practical experience with neurofeedback training, it is my belief that it is a very powerful tool for disciplining attention and developing the conscious mind. Although there are many types of neurofeedback training, I want to emphasize, in particular, alpha and gamma wave training and combinations thereof. Alpha wave training tends to quiet the thinking mind and increase the general flow of energy in both the brain and the body. On the other hand, frontal gamma wave training is unique in raising the energy to higher centres in the brain (the frontal lobes), thereby creating an intense focus. You could say that while alpha training dampens the activity in the old part of the brain (parietal areas), gamma training increases arousal in the new part of the brain (the frontal lobes).

I also want to add that in people with active *kundalini*, it is extremely important to focus the released energy in order to keep it under conscious control. This is best done either through meditation or through alpha and gamma wave training.

TOWARD A HIGHER STATE: CREATIVITY, FLOW AND INTUITION

It seems that instinct and intuition have something in common in that they both initiate spontaneous, appropriate behaviour. You could say that when both work properly, you are in a flow state with no resistance to anything and your behaviour becomes smooth and efficient.

There are of course important differences between instinct and intuition. *Instinct* belongs to the physical realm, while *intuition* belongs to the mental realm. When the body functions in a proper spontaneous way, we call it *instinct*, and when the mind works in a spontaneously perfect way, we call it *intuition*. Both are important, because without instinct you would not be alive, and without intuition your life would have no meaning.

Instinct is physical and relates to the past. It is almost infallible, since it is based on millions of years of evolution. It is seated in the deep brain structures such as the brainstem and the hypothalamus, which constitute the reptilian brain. Instinct is necessary for the survival of both animals and humans. It takes care of the basic needs associated with reproduction and self-preservation. This includes the regulation of the heart, circulatory system, respiration, food intake and sexual activity.

Between instinct and intuition is the rational mind, the intellect. It is a relatively new construct in evolution. It is probably based in the left hemisphere, as this is preoccupied with logical and verbal-analytical thinking. The right hemisphere, on the other hand, being holistic, integrative and able to process several inputs simultaneously, is most likely the seat of intuition.

I doubt that intuition can ever be explained rationally. It is beyond instinct but certainly also beyond intellect, and if it is beyond intellect, how can the intellect ever explain it? However, even if we cannot explain what intuition is in rational terms, we may try to improve our understanding of it in terms of brain function. Then we may be able to develop methods for training it.

Intuition means internal tuition. Usually we acquire knowledge from the outside, which is then processed by the intellect. However, in case of intuition, knowledge comes from the inside as a sudden insight or a knowing without any interference from the intellect. The intellect gives us *knowledge*, while intuition provides us with *knowing*. There is an important difference: knowledge is based on others' experience, while knowing is based on our own experience. Most people have at some point experienced intuition, perhaps in the form of a flash of insight into a certain problem.

I believe that the intellect, with its often incessant thinking, is a barrier to intuition. The best (intuitive) ideas often appear when we are relaxing: in bed, in the shower, during a walk or during meditation. Osho says, 'Intuition opens its doors through meditation. Meditation is simply a knocking on the doors of intuition.' (Osho 2001: 10)

During meditation there is a quieting of the mind's activity. This activity mostly goes on in the left hemisphere and often takes the form of compulsive thinking. It is very likely that busy thinking in the left hemisphere can suppress right-hemisphere activity and thus block intuition. The right hemisphere is the seat of subconscious processes such as imagery, dreaming, fantasies and feelings, so if you can open up to it, you are much more likely to be intuitive, and therefore creative. It seems to me that all creativity must in fact be intuitive. The intellect cannot be creative, since it operates on logic and mechanics. Thus the left hemisphere needs to co-operate with the right hemisphere in order to access intuition and creativity.

Recently a bestselling book came out by Daniel H. Pink entitled *A Whole New Mind*. It had the subtitle 'Why Right Brainers Will Rule the Future'. In the foreword Pink writes:

> *We are moving from an economy and a society built on the logical, linear, computer-like capabilities of the Information Age to an economy and a society built on the inventive, empathic, big picture capabilities of what's rising in its place, the Conceptual Age... Thanks to an array of forces – material abundance ... globalization ... and powerful technologies ... we are entering a new age.*
>
> *(Pink 2005: 2)*

Pink suggests that what is needed in this new age are skills such as the capacity to detect patterns and opportunities, to combine old material into new ideas, to empathize with others and detect the subtleties in non-verbal communication. He goes on to relate these skills to the cerebral hemispheres:

> *The well-established differences between the two hemispheres of the brain yield a powerful metaphor for interpreting our present and guiding our future. Today the defining skills of the previous era – the 'left-brain' capabilities that powered the Information Age – are necessary but no longer sufficient. And the capabilities we once disdained or thought frivolous – the 'right-brain' qualities of inventiveness, empathy, joyfulness, and meaning – increasingly will determine who flourishes and who flounders.*
>
> *(Ibid: 3)*

Split-brain researcher Joseph Bogen suggests that an overly active left hemisphere may block right-hemisphere activity via the *corpus callosum* and prevent access to that hemisphere's vast potential of creativity:

> *Creativity not only made the human species unique in Nature; what is more important for the individual, it gives value and purpose to human existence. Creativity requires more than technical skills and logical thought; it also needs the cultivation and collaboration of the appositional mind [in the right hemisphere]. If the constraint of an intellectual ideal can make man a unilateral being, physiologically underdeveloped, a better informed and foresighted community will strive toward a more harmonious development of the organism by assuring an appropriate training and a greater consideration for the other side of the brain.*
>
> *(Bogen and Bogen 1969: 220)*

It has been suggested that the creative process proceeds in four different phases: preparation, incubation, illumination and verification. Joseph Bogen has suggested an interesting model of brain function that facilitates the creative process: 1) preparation (reading books, formulating hypothesis, asking questions, etc.) goes on primarily in the left hemisphere; 2) the incubation (a simmering or restructuring of the material in the subconscious) takes place in the right hemisphere; 3) illumination comes when you get a bright creative idea, which may be communicated from the right to the left hemisphere via the *corpus callosum*; 4) finally it is up to the left hemisphere to verify the validity and practicality of the idea.

Interestingly, recent EEG research supports the idea that intuitive insights originate in the right hemisphere. Researchers Bavin Sheth, Simone Sandkühler and Joydeep Bhattacharya, in an experiment on problem solving, found increased fast EEG gamma activity in the right frontal lobe up to eight seconds before the subject gained insight into a difficult cognitive problem. It seemed as if the insight originated in the right frontal lobe and only seconds later was conveyed via the *corpus callosum* to the left hemisphere, which interpreted it as an 'aha experience' and a solution to the problem. (Sheth, Sandkühler, Bhattacharya 2009)

A recent study in Australia points in the same direction. Researchers Richard Chi and Allan Snyder from the University of Sydney found that the right temporal lobe seemed to be associated with insight or novel meaning. They also found that inhibition of the left temporal lobe (caused by electrical stimulation) could induce a cognitive style that was less top down, less influenced by preconceptions. (Chi and Snyder 2011)

Maybe the most crucial part of the creative process is the ability to communicate between the two brain halves via the *corpus callosum*. We know that most of the activity in the *corpus callosum* is inhibitory. However, in the creative process we do not want the two hemispheres to inhibit or oppose each other but to co-operate.

In a split-brain patient it is impossible for the two hemispheres to co-operate, since they have been disconnected. Using a number of research methods, Klaus D. Hoppe (1988) found a distinct lack of creativity in these patients. He says that a disruption of the communication between the two hemispheres in split-brain patients creates an 'outstanding lack of creativity' which could be demonstrated using both psychological and EEG methods.

It was found that the disconnection of the *corpus callosum* in split-brain patients decreased EEG alpha band coherence between the left and the right hemispheres. The alpha brain waves in the left and right brain were not in synch anymore. On the contrary, in very creative persons, alpha waves in the left and the right hemisphere were highly synchronized and coherent.

In the same study it was also found that people who had great difficulty in expressing feelings and fantasies (called alexithymia) also showed low coherence of alpha waves between the two hemispheres. Thus Klaus Hoppe (ibid.) concludes that 'Creativity can be understood as the opposite of alexithymia.'

These studies reflect the possibility of measuring (by means of EEG coherence) a very important aspect of creativity, namely the ability for the two hemispheres to communicate via the *corpus callosum* and synthesize information from both sides of the brain. I believe that innovation and creativity emerge on the border between order and chaos, between analysis and intuition, between the conscious and the unconscious and between the left and the right brain.

In our work with neurofeedback we have developed a method that alternately activates the left and the right hemispheres. This is a method of training we call 'in and out of alpha' and it is designed to stimulate the brain to come up with creative ideas. In a trained subject who can maintain high alpha activity for a period of time, we alternate, at two-minute intervals, between alpha feedback with closed eyes and no feedback with open eyes. During the open-eyes period the subject formulates a problem to which they would like to have a creative solution. They may analyse the problem and think about it as much as they want. During this period they use their left hemisphere and their alpha activity will be low. After two minutes they are asked to let go of thinking and close their eyes. Now the alpha feedback is switched on and they are encouraged to keep the alpha tone going continuously for the next two minutes. During this period of time their alpha activity increases and consequently the activity in their left hemisphere is strongly diminished. This gives their right hemisphere a greater chance to express itself. This cycle of 'alpha on' and 'alpha off' is repeated up to 10 times (over 40 minutes).

We believe that turning a subject's alpha activity on and off deliberately at certain intervals makes them alternate between their left and right hemispheres and thus between conscious and subconscious processing. In fact, every time you go into alpha, the 'thinking computer' of the left hemisphere is reset, and when you come out of alpha you can start thinking fresh thoughts. This method prevents you from getting stuck in your thinking.

NN, a CEO of an advertising company in Copenhagen, had done alpha wave training about 10 times before he started training 'in and out of alpha'. After the training he

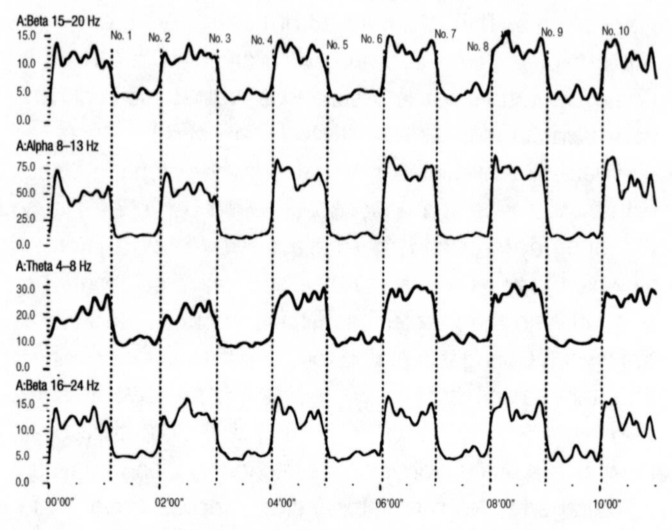

Figure 5.4: Creativity Training: 'In and Out of Alpha'

said, 'Sometimes you have a problem you cannot crack – you are stuck in your thinking. The next morning when you wake up, the solution is there all of a sudden… When I train "in and out of alpha", I simply cheat the brain to believe it has slept several times in 20 minutes.' NN claimed that during the training he had generated several creative ideas, which were later successfully applied in his work with advertising.

I discussed this training method with Dr James Hardt, head of the Biocybernaut Centre in California. He explained that their alpha wave training was also based on two-minute intervals. Following each interval, the result of the training (measured in microvolts) was fed back to the subject on a display for eight seconds before they went on to the next alpha training interval. Jim also believed that this form of training could stimulate creativity and transcendent experiences. He referred me to a book by Michael Murphy describing spiritual experiences on the golf course. It seems that when people play golf they alternate between intense concentration (on their swings and putts) and relaxation (in between swings). In fact, playing golf may have an effect that is somewhat similar to the 'in and out of alpha' training.

Debbie Crews at the University of Arizona has done an interesting study of how the brain works when a person is playing golf under stress (Crews and Landers 1993). All the golfers in the study subjectively reported increases in anxiety when stressed and showed increased heart rates when stressed. The difference between those who played successfully under stress and those who didn't was reflected in the balance between the left and the right hemisphere, as seen on brain scans. The players who did well had a good balance between the two hemispheres. It seemed that they were able to quiet the left hemisphere's thinking activity under pressure, thus avoiding blocking the right side of the brain. On the other hand, the players who choked under stress showed very high activity in the left hemisphere and very little activity in the right. They were probably analysing their game too much because of the pressure. Following the stroke in her left hemisphere, the neurologist Jill Taylor also had something to say about the importance of the right hemisphere in creativity:

> *I have found that often the last thing a really dominating left hemisphere wants is to share its limited cranial*

> *space with an open-minded right counterpart. My right mind is open to new possibilities and thinks out of the box. It is not limited by the rules and regulations established by my left mind that created that box. Consequently, my right mind is highly creative in its willingness to try something new. It appreciates that chaos is the first step in the creative process.*
>
> *(Taylor 2008: 140)*

Increased access to the right hemisphere seems to be accompanied by increased alpha activity. If you access the deep unconscious seated in the limbic system, first alpha waves and then theta waves will appear in the cortical EEG. This is what happens during deep meditation. Theta waves reflect unconscious activity and have been associated with creativity.

Biofeedback pioneer Dr Elmer Green at the Menninger Foundation in Topeka, Kansas, was the first to explore the creative possibilities of alpha/theta training. I was myself a subject in his laboratory in 1971 and tried his 'psycho-physiological training for creativity'. When you go into these deep states, either through meditation or through brainwave training, both alpha and theta waves increase and spread to the whole brain. In these states you often oscillate between alpha and theta, between the conscious and the subconscious. It is sort of a dreamy state often characterized by vivid imagery from the subconscious. It was in such a state that the chemist Kekulé discovered the structure of the benzene ring – a major discovery in chemistry.

A group of English researchers at the Royal College of Music in London did a neurofeedback study in order to see if alpha/theta-training could improve musical ability and creativity in students (Egner and Gruzelier 2003). The results, which were replicated across two years of research, showed that students who had just ten 15-minute alpha/theta sessions over six weeks improved their overall musical performance and interpretative imagination by about 15 per cent, while a control group did not improve. The researchers suggest that the alpha/theta-training involves a transient entering and re-entering of dream-like states (theta) while staying awake (alpha). This procedure supposedly enables consciousness (left hemisphere) to access the subconscious memory of the right hemisphere, leading to more creativity.

I believe that both intuition and creativity are associated with the flow state. This is when we are totally immersed in an activity (physical or mental) and the mind-body system works efficiently without any resistance. It has been suggested (Csikszentmihalyi 1991) that the flow state is a kind of trance. That seems paradoxical, since in a flow state a person is intensely focused and highly effective. On the other hand, they are also totally immersed in the task at hand, forgetting both time and place and ignoring all stimuli unrelated to the task. This could reflect a trance-like state. Since the flow state is trance-like and highly alert and focused at the same time, I sometimes call it a *conscious trance*.

I would like to suggest a brain model of the flow state, which is related to the model of brain awakening discussed elsewhere in this book. I believe that in the flow state, the parietal lobes (the association and thinking areas of the brain) partially shut down (that would explain the trance), while the frontal cortex and right hemisphere are activated (which would explain the intense focus and also the intuitive flow of behaviour). Thus I see the flow state as akin to a higher state of consciousness.

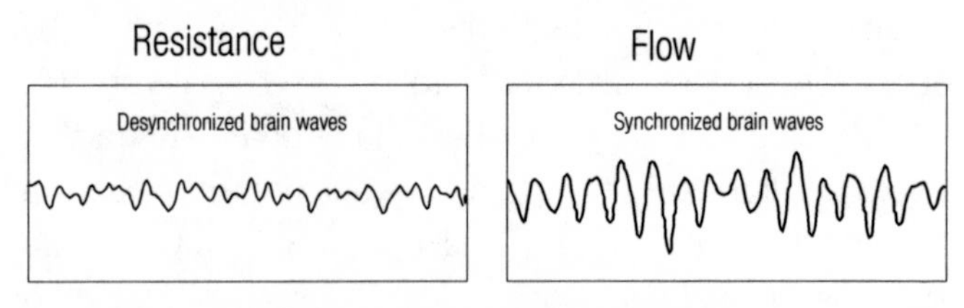

Figure 5.5: Flow versus Resistance

If you give up all resistance and stay in the moment, you will be in a flow state, totally relaxed and yet very efficient in whatever you are doing. Flow states have their own special brainwave signatures, which are well-synchronized high-amplitude brain waves. If the flow is in the body, the brain produces alpha waves; if the flow is in the mental realm, the brainwave signature is fast beta and gamma frequencies, predominantly in the frontal area. Synchronized high-amplitude brain waves indicate that the neurons in the brain are working together; they are coherent and coordinated. The moment you change your focus or resist something, mentally or emotionally, this neural coherence breaks up and the brain waves desynchronize, with smaller amplitudes and faster frequencies.

When we train frontal gamma, people often go into a conscious trance. They say they feel kind of sleepy, but at the same time they are very alert and respond immediately to stimulation. When people start habitual thinking, their gamma level decreases abruptly, probably because the frontal lobes are deactivated and the parietal association areas are reactivated. They lose their focus and their thinking becomes unobserved and automatic (*see Figure 5.3*).

In some cases, however, the subject can maintain a high gamma level while thinking. In these cases the thinking seems to be conscious. It is initiated at will and all thoughts are observed consciously. One client of ours, a research psychologist, had a high frontal gamma level initially and was able to increase it further using neurofeedback. Interestingly, he was able to sustain it while thinking about complicated statistical problems related to his research. Obviously he was intensely focused on his scientific problem and all his thoughts seemed to be conscious.

After the training, he explained: 'In the high gamma state my mind was crystal clear. I had a great overview and penetrating focus. In the gamma state I experienced a greater ability to see and understand and acquire new insights and creative solutions to my problems.'

Unconscious thinking is probably what reduces the frontal gamma level.

To sum up, I believe that all creativity is intuitive. The mechanistic intellect in the left hemisphere can hardly be creative without input from the holistic right hemisphere. On the other hand, we need the left hemisphere with its intellect to communicate the creative ideas to the world. In order to be a whole creative being we need the co-operation of both hemispheres, of both intellect and intuition. In addition, we need the support of strong and active frontal lobes. They direct attention and sustain focus and must therefore play a very important role in creativity.

Since creativity is intuitive and intuition is based on flow, these three concepts are closely interrelated and will all be an important part of the new consciousness.

Chapter 6

A NEW BRAIN AND A NEW CONSCIOUSNESS

Is there a special area in the brain responsible for religious feelings and spiritual experiences? Is there a 'God spot' in the brain? Recent brain research seems to have come up with ambiguous results regarding the issue. Michael Persinger of Laurentien University in Ontario, Canada, has tried to create religious feelings in normal humans artificially by stimulating their brains with weak electromagnetic fields. Persinger and his team have tested hundreds of people with a device he calls the 'God helmet'. Most of these subjects experienced a 'sensed presence' or the presence of a 'sentient being' in the room when there was no one there. States of cosmic bliss were also reported by some, while others felt nothing or even a little anxious or uncomfortable. The electromagnetic stimulation seemed to be most effective over the right temporal lobe. This is why this area in the brain has been termed the 'God spot' by some.

It may be that when the right hemisphere is activated by the electromagnetic stimulation, the left hemisphere perceives that stimulation (via the *corpus callosum*) as someone being there. Thus the activation of the right hemisphere is sensed by the left hemisphere as the presence of another 'sentient being', and this is what Persinger (2003) calls a prototype of the God experience.

It has been suggested that Persinger's God helmet induces electrophysiological changes in the temporal lobes that are somewhat similar to an epileptic seizure, although much more subtle. It has been known throughout history that epileptic seizures may be accompanied by spiritual experiences, such as feeling the presence of God. That seems to have been the case in historical persons such as Dostoyevsky, Saint Paul, Saint Teresa of Avila and others.

Some neuroscientists have suggested that the limbic system is a key brain structure responsible for religious and spiritual experiences. They speculate that during an epileptic seizure in the temporal lobes (which are heavily connected to the limbic system) there is an electrical 'shortcut' between the temporal lobes and limbic system, which may spark religious feelings. There may be some truth to that theory; however, it does not explain away religious/spiritual experiences as brain artefacts (brain anomalies). Rather, I suggest that these experiences are real, they belong to a reality different from our normal daily reality, and they can be experienced only under certain conditions of the brain.

Persinger, incidentally, is not a religious man himself. On the basis of his results, he argues that religious feelings and belief in God *are* just brain artefacts. He even calls belief in God a 'cognitive virus'.

To return to the God spot in the brain, using brain scans US professors Richard Davidson and Andrew Newberg both found increased brain activity in the frontal lobes during spiritual experiences in the deep meditative state (Newberg, D'Aquili 2001; Lutz *et al.* 2004).

In my mind, the frontal lobes and the right hemisphere are crucial for experiencing higher states of consciousness. When Jill Taylor experienced reality through her right hemisphere only, she described it as being:

> *...open to the eternal flow whereby I exist at one with the universe. It is the seat of my divine mind, the knower, the wise woman, and the observer. It is my intuition and higher consciousness.*
>
> *(Taylor 2008: 140)*

From the above research it seems that if there is a God spot in the brain it is either in the temporal or the frontal lobes, but is also likely to involve the limbic system and the right hemisphere. It seems that several brain structures are involved in spiritual feelings and the experience of God.

This is also the opinion of neuroscientist Mario Beauregard of the University of Montreal. Using MRI brain scans, he studied 15 Catholic Carmelite nuns who claimed to have had an experience of intense union with God. When they were experiencing communion with God, or at least having a deep spiritual experience, their brains were activated in no fewer than six different places, including areas in the frontal lobes (Beauregard, O'Leary 2007).

Thus we are led to conclude that there is not one single God spot in the brain but that neural networks distributed throughout the brain mediate spiritual states.

THE AWAKENING OF A NEW BRAIN

Andrew Newberg, professor of nuclear medicine at the University of Pennsylvania and author of the acclaimed book *Why God Won't Go Away*, studied eight Tibetan Buddhist meditators using SPECT scans. The images he captured showed that during deep meditation there was an increase in blood flow and neural activity in the prefrontal cortex, while at the same time, surprisingly, a sudden drop in activity in the parietal area, the upper rear part of the brain, which Newberg calls the orientation association area (OAA). This shutting down of the OAA was especially pronounced on the left side, where the thinking mind is situated.

Newberg theorizes that when the meditator withdraws from the outside world, sensory inputs to the parietal areas (OAA) are blocked and most neural activity in that area is shut down. At the same time, due to the intense concentration, the prefrontal cortex, or attention association area (AAA), is strongly activated (Newberg, D'Aquili 2001).

The OAA on the left side is the area that gives us the ability to orient ourselves in time and gives our body a sense of physical limits and our self a sense of separateness from the rest of the universe. Newberg argues that when the parietal areas in the cortex are deactivated, the physical limits of the body and the sense of separateness disappear. The brain can no longer create a boundary between the self and the outside world, or locate itself in physical reality. As a result, Newberg says, it has no choice but to perceive itself as endless, interwoven with everyone and everything. This is the state Newberg calls *Absolute Unitary Being*. He

AWAKENING DEPENDS ON THE SHIFT IN A BRAIN FUNCTION – FROM PARIETAL (OAA) TO FRONTAL (AAA) AREA

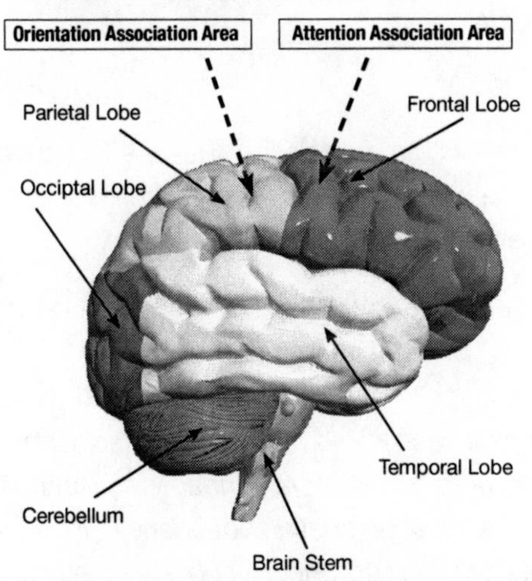

Frontal Lobe (AAA):

- Willpower
- Focus of attention/energy
- Experience of Oneness

Parietal Lobe (OAA):

- Orientation in space
- Distinction between the Self and the world
- Seat of Mind and Ego

Figure 6.1: Awakening Depends on a Shift in Brain Function from the Pariental (OAA) to the Frontal (AAA) Area

suggests that people in this state experience a primary reality of oneness with everything. He is certain that we have evolved a capacity to perceive an absolute, universal reality in which we are all fundamentally connected as one.

Newberg's theory has been criticized by physicist Peter Russell for being too simple and mechanistic:

> *...to equate the dissolution of the self with a decrease in the brain processes that govern our orientation in space, hypothesizing that this leads to a corresponding loss of boundaries between the self and the rest of the world, misses the real nature of this state.*
>
> *(Russell 2003)*

I suggest that the correlation between reduced brain activity in the left parietal cortex and the dissolution of self may still be valid, even if it is a crude theory.

Russell elaborates on the real nature of the dissolution of the self:

> *For experienced meditators, the dissolution of the self is something far more fundamental and significant than a loss of spatial boundaries. In these deep states, there is no longer a subject–object relationship to experience; no longer a separate 'I' observing experience or thinking thoughts. Experience happens, just as before. Thoughts may still arise in the mind. But, paradoxical as it may sound, there is no one thinking them... The individual sense of self that is so*

> *familiar, a seeming part of every experience, is seen to be but a construct in consciousness, another experience arising in the mind.*
>
> *(Ibid.)*

Newberg's research with Iversen (2003) has led him to conclude that the right prefrontal cortex plays a crucial part in meditation. His brain model of meditation begins with activation of the prefrontal cortex, particularly in the right hemisphere. (Newberg and Iversen 2003).

In another study using positron emission tomography (PET) scans and quantitative electroencephalography, a research team in Oklahoma City studied an Indian master, Swami Nithyananda, while he was going into deep meditation. There were two remarkable findings:

> *First, the dominant hemisphere of Swami's brain was more than 90 percent shut down... A second amazing aspect of Swami's deep meditation was that the lower portion of his mesial frontal areas lighted up in a very significant way... When we later asked Swami what he was doing when the mesial frontal areas lighted up, he said he was opening his third eye.*
>
> *(Murali Krishna, www.dhyanapeetam.org/web/Oklahoma_Research_Report.htm)*

The strong decrease of left-hemisphere activity found in this study during deep meditation is significant, I believe, and is somewhat similar to my own findings in the ayahuasca study. Together with the findings of Richard Davidson and

Andrew Newberg (of strongly increased frontal lobe activity during deep meditation in experienced Buddhist monks), it points to radical brain changes during higher states of consciousness.

The results of my own and others' research have led me to formulate a hypothetical model for future brain evolution. I foresee that a fourth level of brain evolution (on top of the Triune Brain) will include more energized and active frontal lobes together with an opening up of the *corpus callosum*. This will lead to increased activity in the right hemisphere, which eventually will become better integrated with the left and an equal partner with it. This is what I call the *New Brain*.

This theory implies that the process of awakening is not merely due to psychological changes but primarily due to a fundamental change in brain function, with a shift in brain dominance from the parietal to the frontal areas and from the left to the right brain. When the over-activity in the parietal areas is decreased and the under-activity in the frontal areas is increased, there is a shift in the brain's activity pattern and the frontal cortex takes total charge of brain functioning.

The reduction of brain activity in the parietal areas, as found by Newberg during deep meditation, most likely corresponds to a decrease of mind (and thinking) activity. I refer to that part of the brain as the 'Old Brain'. In the New Brain, on the other hand, the increased activity in the frontal cortex and in the right hemisphere, I believe, corresponds to more awareness, presence and intuition as part of a general heightening of consciousness.

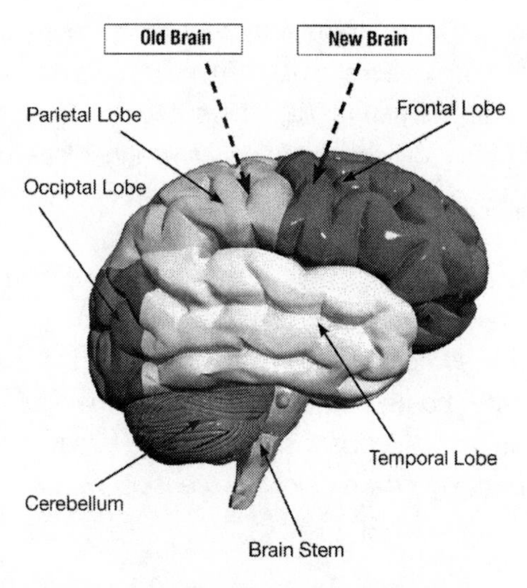

Figure 6.2: The 'Old Brain' and the 'New Brain'

A HIGHER LEVEL OF REALITY

It seems that when a shift occurs from the Old to the New Brain (from the parietal to the frontal lobes), the individual wakes up to a higher level of consciousness and to a new reality which, Newberg argues, 'is even more real than the old one'.

The only way one can judge the reality of an experience is by how real it feels. When you are in a dream it feels real, but when you wake up, normal daily reality feels much more real. The same is the case when you wake up to a higher level of consciousness: all experiences are bound to feel more real than ever.

Every morning when you wake up from sleep, you wake up to a higher level of consciousness. Actually, you wake up twice: first you may wake from deep sleep to dream sleep and later you wake from dreaming to your normal daily reality. What I am suggesting here is that it is possible to wake from that daily reality to an even higher level of reality.

Gopi Krishna puts it that way:

> *Just as a man awakening from a frightful dream returns to his normal state with a deep sense of relief and gratitude, in the same way the awakened seeker contemplates, with joy and thankfulness, the newly transcendent state of being, released from the cramping prison-cell of the sensory world.*
>
> *(Krishna 1975: 59)*

There is a general claim by Eastern masters that at the present level of evolution we are still in sort of a dreaming state from which we need to wake up. This theme has also been treated by several Western authors: Charles Tart in *Waking Up*, Peter Russell in *Waking Up in Time* and most recently Steve Taylor in *Waking from Sleep*.

In order for the human brain to wake up and attain a higher state of consciousness, functional and biological changes of the brain are needed. I believe that evolution's target for a new human brain involves a more mature and active frontal cortex together with a more developed right hemisphere better integrated with the left. This is the awakened brain, the New Brain, which can give rise to spiritual experiences of the kind Gopi Krishna describes.

As we have already seen, each level of consciousness (and its associated reality) is based on a specific brain structure and reflected by a specific brainwave frequency. At the lowest level is the reptilian brain, with its survival functions reflected by slow delta waves. At a somewhat higher level is the mammalian, emotional brain, with the limbic system generating theta waves, as seen for instance during strong emotions and during dreams. These systems are active when you sleep or dream and they let you experience their corresponding realities. When we wake up in the morning the neo-cortex is activated and the brain starts producing alpha and beta waves, enabling us to access what we call normal reality. What I am proposing here is that when the frontal area of the brain is energized through proper evolutionary maturation and the right hemisphere activated as a consequence, we wake up to a higher level of reality.

Before, it may be that we were only looking at the world via our left hemisphere's mechanistic system, and that world may have appeared shadowy, dull and sterile. After awakening, however, we can look through our whole brain (left and right hemisphere plus the emotional limbic system) and the world will appear much more radiant, real and alive. This, of course, is also because we, ourselves, have become more real and alive.

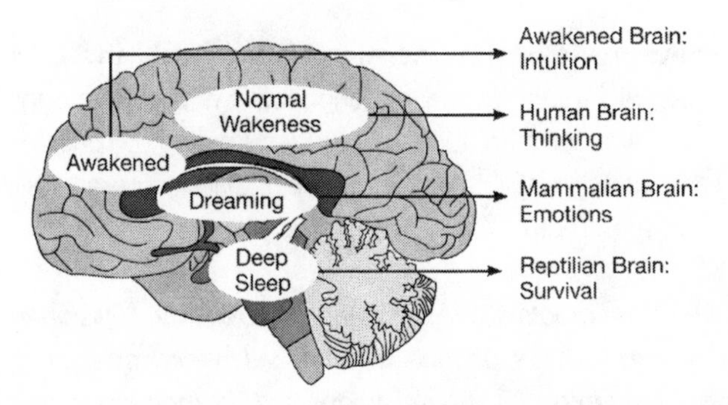

Figure 6.3: Brain Structures Associated with Different Levels of Consciousness

Altered States versus Awakening

Altered states are sometimes confused with higher states of consciousness, and mystical experiences are confused with spiritual (awakening) experiences. In my definition, altered states belong to the unconscious and subconscious realms and are associated with the activation of deep brain structures including the limbic system and the right cerebral hemisphere. That is why I also call lower states or deep states of consciousness 'altered states'. An altered state does not qualify as an awakened state or a higher state of consciousness. In an altered state, deep brain structures and the right hemisphere seem to be involved, while the frontal lobes are only affected to a smaller degree. In most (if not in all) altered states of consciousness I believe that it is the activation of the right hemisphere that gives rise to mystical experiences. In these states there seems to be no activation or even inhibition of the frontal lobes and the whole left hemisphere. Spiritual experiences during awakening, however, seem primarily to activate the frontal lobes and only secondarily involve the right hemisphere.

I believe that Gopi Krishna was thinking about this issue – mystical versus spiritual experiences – when he strongly warned against meditative techniques leading to 'passive, quiescent, vacant or semi-awake conditions represented by Alpha or Theta waves'. He wrote:

> *The practitioners, coming across vivid visionary experiences or face to face with weird spectacles presented by the subconscious, in their ignorance,*

> *often accept them as signs of spiritual awakening, which actually is far from the case.*
>
> *(Krishna 1975: 56)*

Krishna repeatedly emphasized that the necessary prerequisite for spiritual awakening was 'the power of sustained attention cultivated by voluntary effort'. Since sustained attention is mediated by the frontal lobes, these ideas are in accordance with my own concept of frontal cortical activation as a necessary precondition for awakening.

When people have mystical experiences involving fantastic visions through the use of meditation, drugs, biofeedback or other mind-altering techniques, they are *not* in a higher state of consciousness but rather in a lower state confronting the subconscious mind, experiencing a reality akin to dreaming, a state which is likely to be associated with increased right-hemisphere activity. Mystical and spiritual experiences are thus principally different subjectively as well as objectively, since they are based on different brain structures.

People who use marijuana, psilocybin, mescaline or ayahuasca may experience very entrancing visions but they also experience a loss of intellect and judgement. But in the case of a genuine spiritual experience, 'all normal faculties of mind are enhanced, not blunted', according to Gopi Krishna, who continues:

> *In the higher state of consciousness you can see the real world even more real. You can see the same things magnified, without any visions or strange creatures*

or projections. You see the same thing expanded, and you see yourself by means of this expansion, one with the creation around you – without the least loss of precision of intellect, or sight, or smell.

(Ibid: 123)

In another interview Krishna says:

In a genuine mystical experience, there is clarity, lucidity. In fact, the clarity and lucidity are a hundred times more powerful than in normal consciousness. It is not like a dream or a drug experience.

(Quoted Kieffer 1983).

Unfortunately Gopi Krishna uses the words 'mystical experience' synonymously with a spiritual or awakening experience. I prefer to use 'mystical experience' in reference to altered states, since these experiences are often mystical, strange and weird, while the genuine spiritual experience is very lucid and very real.

THE NEW BRAIN AND THE ONENESS STATE

The New Brain, with its more energized frontal lobes, together with an active and better integrated right hemisphere, causes profound changes in perception. As already noted, Andrew Newberg theorizes that when the frontal cortex is activated and the parietal areas somewhat deactivated (as is the case during certain types of meditation), the subject experiences a state of

Absolute Unitary Being – a primary reality of oneness with the universe and everything in it. This has been called the *Oneness State* by the Oneness-Movement and its founder in India, Bhaghavan Kalki.

In a world of people with a New Brain there would be no feeling of separation from other people or from the universe. Everyone would feel connected with everyone. As the part of the Old Brain concerned with temporal sequencing would have become very quiet, people would lose interest in both the past and the future. All their focus would be on the present moment. But what are the relations between past, present and future? Let's turn our attention to the puzzle of time.

From Time to Presence: The Dissolution of the Ego

The ego is very much preoccupied with time, especially with the past and the future. According to Eckhart Tolle, this is because 'the past gives you an identity and the future holds the promise of salvation, of fulfilment in whatever form'. He says:

> *End the delusion of time. Time and mind are inseparable. Remove time from the mind and it stops... To be identified with the mind is to be trapped in time: the compulsion to live almost exclusively through memory and anticipation. This creates an endless preoccupation with past and future.*
>
> *(Tolle 1999: 40)*

Osho claims that the only reality is in the present moment. He talks about three different types of time:

> *Chronological time corresponds to the body, psychological time to the mind, and real time to your being. Chronological time is the extroverted mind, psychological time is the introverted mind and real time is no-mind … real time is not a process. It is simultaneity. Future, past, present are not three separate things … it is an eternal now, it is eternity. And finally: future and past are mental states, not part of time, time is eternal.*
>
> *(Osho 2005: 191–9)*

Thus the present moment is all there is – the only reality. The eternal present is the space within which all life unfolds.

And according to my model, as the New Brain develops, the ego dissolves as the activity shifts from the left to the right hemisphere. This shift causes less preoccupation with time and more focus on the present moment. This is what is called *presence*.

Eckhart Tolle says, 'Presence means consciousness becoming conscious of itself.' (Tolle 1999: 81) In terms of neurology, it may have something to do with the interaction between the cerebral hemispheres. It is hard to conceive that in isolation the left-hemisphere mind can be conscious of itself or that the right-hemisphere mind can be conscious of itself. However, it is quite conceivable that either hemisphere can be conscious of the other (because they are connected via the *corpus callosum*) and that this is a necessary condition for being self-conscious, self-reflective and present. Thus I propose that presence is when each hemisphere is conscious of its counterpart, when there is direct contact and communication between the left and the right sides of the brain.

Unity in Experience and Unity with Cosmos

The perception of time seems to be associated with the left hemisphere, where a sequencing mechanism maps time moment by moment. Contemporary neuroscientists have begun to wonder whether our perception of the world is continuous or is a series of discrete snapshots like frames on a film strip.

A French neuroscientist, Rufin VanRullen, has found that the brain's receptivity to new visual information 'correlates with the oscillatory phase of the ongoing EEG brainwaves'. Experiments showed that receptivity to visual stimulation shifted up and down in accordance with the phase of the EEG signal, leading to something akin to discrete visual frames. VanRullen says: 'There's a succession of "on" periods and "off" periods of perception… Attention is collecting information through snapshots.' (Busch, Dubois, VanRullen 2009: 7,869)

It is interesting that in Buddhist thinking the stream of consciousness consists of successive moments of cognition. According to this philosophy, a single moment of attention is focused on only one of the five physical senses at a time, or on the thinking mind. If the attention switches quickly enough between the senses (multiplexing), all the inputs will appear simultaneous and merge into a coherent experience. This would work similar to a movie where 24 separate frames are shown per second, resulting in the experience of smooth, realistic movements on the screen.

Consciousness philosophers have long been wondering how all the inputs from the different senses come together in the brain to form a whole experience. This is called the 'binding problem'. Buddhist philosophy seems to offer a solution to that problem.

Jill Taylor gives an illustration of the above theories from her right-hemisphere perspective:

> *Time stood still because that clock that would sit and tick in the back of my left brain, that clock that helped me establish linearity between my thoughts, was now silent... Instead of a continuous flow of experience that could be divided into past, present and future, every moment seemed to exist in perfect isolation.*
>
> *(Taylor 2008: 48 and 50)*

Fast 40 Hz gamma brain waves have been associated with the 'binding' of sensory inputs together in the brain to form a whole experience. If we assume, with VanRullen, that the multiplexing (scanning) frequency correlates with the EEG frequency, it is certain that scanning the sensory inputs at 40 cycles per second would bind the inputs together to one whole experience.

We know that when the brain operates at slow frequencies (below 8 Hz), such as delta and theta, we are unconscious/ asleep, and during conscious states, the brain operates at faster frequencies in the alpha and beta band. As we have seen, there is evidence that in the super-conscious or awakened state the brain operates at very fast gamma wave frequencies, especially in the brain's prefrontal area, where I assume the 'binding' process takes place. One hypothesis would be then that sensory information, as well as thoughts and feelings, is 'updated' from various brain areas to the prefrontal cortex at a speed corresponding to the brainwave frequency. The brain of a normal, awake and conscious person usually operates at between 10 and 20 cycles per second. At very fast brainwave frequencies, such as the 40 Hz gamma, the scanning of the sensory inputs will be so fast that it will generate high-resolution perception. That may be why an awakened person experiences the world in HD. This model also accounts for an awakened person's ability to perceive even the smallest details in the environment.

As already noted, when the EEG frequency drops below 8 Hz, we are asleep. However, *what would happen if it were possible to lower the brainwave frequency below 8 Hz without falling asleep?* (This is actually what may happen during an ayahuasca trip when the ego is falling apart.) Indian master Bhagavan Kalki offers an answer:

> *...the senses are coordinating at a particular speed whereby you are feeling that you are seeing, touching as though all these things are going on at the same time. It is this illusion that creates the sense of 'Me' and 'I' [the ego]. If this is slowed down, you all vanish... But you will still be very functional, you will in fact be very efficient. The sense of separateness would completely be gone... It's a simple thing – the senses*

> *have to be slowed down a little bit – and behold; you are not there. What is there is only life. You will feel you are one with the universe, completely one.*
>
> *(Quoted Windrider www.onenessforall.com)*

So, in order to maintain focus and perceive unity in experience, the frontal lobes must be running brainwave frequencies above 8 Hz. But in order to perceive unity with cosmos and everything in it, I suggest that they must be running much faster brainwave frequencies in the 40 Hz gamma range. At the same time, the Old Brain (the parietal lobes) must slow down its frequencies, probably below 8 Hz, and also shift its activity toward the right hemisphere. I suggest these are the necessary conditions for a New Brain person to experience the oneness state.

A state of absolute knowingness and feeling at one with the universe has been reported by mystics for millennia, and it seems to be associated with the right hemisphere. Jill Taylor, the neurologist who had a stroke in her left temporal lobe, observed from her right hemisphere this state of oneness with the universe:

> *As the language centers in my left hemisphere grew increasingly silent and I became detached from the memories of my life, I was comforted by an expanding sense of grace. In this void of higher cognition and details pertaining to my normal life, my consciousness soared into an all-knowingness, a 'being at one' with the universe, if you will.*
>
> *(Taylor 2008: 41)*

THE RIGHT HEMISPHERE IN EVOLUTION

In my theory of the New Brain, the right hemisphere plays a major role. It seems that at the present level of evolution most right-brain activity is unconscious or subconscious. This is partly because the right brain is underdeveloped and partly because it is inhibited by the left, dominant hemisphere via the *corpus callosum*. However, I believe that evolution is headed towards a more active right hemisphere in humans that is better integrated with the left. As this development progresses, the unconscious content of the right hemisphere will be made conscious. I believe that meditation, psychotherapy and the use of mind-expanding substances are all ways to facilitate a shift of activity from the left to the right hemisphere.

Since we have already established that both cerebral hemispheres are conscious at a human level, one might ask if there is any principal difference between left and right-hemisphere consciousness?

Eckhart Tolle distinguishes between what he calls *object consciousness* and *space consciousness*. Intuitively, I tend to relate these two types of consciousness to the left and the right brain hemisphere respectively. Tolle says that most people's lives and minds are filled with material things and cluttered up with thoughts. This is the dimension he calls *object consciousness*. *Space consciousness*, on the other hand, refers to 'an undercurrent of awareness' or a background awareness.

Neurophysiologist Rhawn Joseph referred to the right-hemisphere mind as an 'ancient awareness' and that is exactly what it is – it is an awareness originating from millions of years of phylogenetic development and it represents who we really are. If we can sense that ancient or background awareness in all our doings (including our thinking), we are present. Tolle says:

> *If you can sense an alert inner stillness in the background while things happen in the foreground – that's it! This dimension is there in everyone, but most people are completely unaware of it... Object consciousness needs to be balanced by space consciousness for sanity to return to our planet and for humanity to fulfill its destiny. The arising of space consciousness is the next stage in the evolution of humanity.*
>
> *(Tolle 2005: 227–8)*

If I am not wrong, object consciousness relates to the left brain while space consciousness belongs to the right brain. So what Tolle is actually saying is that we need more right-brain awareness to provide us with a space for all our activities in the material world and to give us a true perspective on all the left-hemisphere mind activities.

It has been assumed (Joseph 1992) that the frontal lobes stand guard on either side of the *corpus callosum* determining what information may pass. It is possible that the frontal part of the *corpus callosum* only opens up when the frontal lobes are sufficiently energized by running fast gamma brain waves. Thus frontal activation (for example following a *kundalini* arousal) may cause the *corpus callosum* to open up so the integration process of the two hemispheres can begin.

Inner and Outer Wars: The Repression of the Right Hemisphere

We cannot bring peace to the world as long as we are at war with ourselves. Almost every one of us has some internal conflict going on; we are not at peace with ourselves. I believe that a great deal of this inner conflict is reflected in fights between the two cerebral hemispheres.

It is said that if conflict between the hemispheres is not solved on the inner level, it will be projected to the outside. So, if we do not resolve our disputes on the inner plane, we will meet them on the outside and be forced to resolve them at the physical level.

The great Indian master Osho intuitively knew that:

> *These two hemispheres are constantly in conflict – the basic politics of the world is within you... You may not be aware of it, but once you become aware, the real thing to be done is somewhere between these two minds.*
>
> *(Osho 2001: 123)*

It is important to be aware of this, especially if you are in a relationship. Osho gives an example:

> *And this is my understanding: unless you have resolved your inner fight between the right and*

> *the left hemispheres, you will never be able to be peacefully in love – never – because the inner fight will be reflected on the outside. If you are fighting inside and you are identified with the left hemisphere, the hemisphere of reason, and you are continuously trying to overpower the right hemisphere, you will try to do the same with the woman you fall in love with.*
>
> *(Ibid: 129)*

This is also called projection: an internal conflict is projected to the outside, where the person tries to solve it. This effort, however, will always be in vain.

The Repression of Sex, Feelings and the Feminine

I am convinced that the general repression of the right hemisphere in humans is an attempt to suppress feelings, sex and the feminine. These important aspects of a human being are deeply rooted in the right hemisphere and its connected limbic system. Therapies, meditation and mind-expanding drugs which affect the right hemisphere all seem to stimulate both feelings and sexuality.

In many civilized societies in modern times sex has been something you would rather not talk about in the family and in society. However, this was not the case in ancient times. There sex was a thing to be celebrated! The genitals were not considered to be obscene, and in some countries they were barely covered. The Greeks, Romans and Egyptians all left depictions of this sexual culture, leaving little to the imagination.

So what happened? Why did sex become 'dirty' and 'unsuitable for dinner conversation'? And why did Adam and Eve cover their genitals with fig leaves? I believe the answer can be found in the evolution from right to left-hemisphere dominance and in the accompanying struggle for power between the hemispheres. In a later chapter I will elaborate on this subject.

Science has already found that in what we call normal people there is, in addition to a certain amount of co-operation between the two brain halves, an ongoing struggle for control and dominance. It has been determined that most of the activity across the *corpus callosum* is inhibitory in nature, reflecting the fact that each hemisphere is trying to dominate the other. It seems to involve active competition between the hemispheres, which the left hemisphere is genetically destined to win.

Repression of sex, feelings and the feminine all reflect the general repression of the right hemisphere. In the ancient world where people, I believe, primarily experienced the world through their right hemisphere (and its closely connected limbic system), sex was a natural thing just like eating. My theory is that as the evolutionary journey from the right to the left hemisphere started, the latter began to reflect on the sexual behaviour of the right hemisphere and to condemn and oppose it, and a power struggle began between the hemispheres. From that point, nakedness, sexuality and the feminine were banned, opposed and persecuted – more so in some countries than in others. In certain parts of the world this trend is very obvious. We see ritual genital mutilation of young boys and girls, capital punishment for sexual deviants and the suppression of women, whose behaviour is restricted and who are often required to cover themselves up completely. In certain societies, when a woman is believed to have brought dishonour to her family through adultery, she may be killed by a father or a brother. Such 'honour killings' are extreme reactions to unwanted sexual behaviour in women.

The sexual liberation and the widespread use of meditation, therapy and mind-expanding drugs we saw in some parts of the world in the sixties may be seen as a revolt against and rebound from too much left-hemisphere activity and an attempt to return to the right hemisphere. This was actually the view of US professor of psychiatry Roland Fischer:

> *We submit that this technological, scientific, rational belief system has resulted in an overemphasis on the logical-analytical cognitive mode of the left hemisphere, and this in turn has resulted in a rebound effect, characterized by a dramatic over-compensatory swing toward the cognitive mode of the right hemisphere. The symptoms of this rebound include a declining interest and participation in organized, pre-structured science and religion and a corresponding upsurge in experiential religious pursuits such as those offered by Eastern meditation, hallucinogenic drug use… This reflects the view that God never died, He has been alive and well in the right hemisphere all the time.*
>
> *(Fischer, Rhead 1974: 197)*

The Universe Within

Some enlightened masters of the East claim that the whole universe is contained within oneself. Physicist Peter Russell reflects over the statement 'I am the universe' coming from spiritual adepts with many years of inner exploration. Are these people crazy or do they have an extraordinary, intuitive insight? Russell writes:

> *It is far more likely that they represent people who have experienced first-hand that the entire universe – everything we know from the cells in our bodies to the distant twinkling stars – exists within the mind, not the other way around.*
>
> *(Russell 2005)*

In an interview with Amy Edelstein, Indian master Mata Amritanandamayi, also known as Ammachi or Amma, 'the hugging saint', talks about the necessity of changing on the inner plane in order to change the world:

> *There is a divine message hidden behind every experience life brings you – both the positive and the negative experiences. Just penetrate beneath the surface and you will receive the message. Nothing comes from the outside: everything is within you. The whole universe is within you.*
>
> *(Amritanandamayi, quoted Edelstein 2000)*

Another Indian master, Bhagavan Kalki, agrees when he says:

The world is a manifestation of our inner state. The situations we come across, the people we meet, the problems we confront, and the varieties of life experiences we have, is a projection of what lies within. In other words we create our reality... You become what you think.

(Kalki, www.onenessuniversity.org.)

I believe that when the right brain is fully integrated with the left we will realize that we are one with all other living beings. What we think of others, we think of ourselves, and when we meet others, we actually meet ourselves.

The hypothesis that the whole universe is represented within oneself, is also supported by Shaivism, the oldest sect of Hinduism, which states that kundalini makes one expand infinitely, assimilating the whole universe into one's self.

I have already argued that the feeling of oneness with God and with the universe is somehow linked to the right brain. Could it be that the whole universe is represented in the right hemisphere?

Neurosurgeon Joseph Bogen philosophizes about hemispheric complementarity: 'Each hemisphere represents the other and the world in complementary mappings; the left mapping the self as a subset of the world and the right mapping the world as a subset of the self.' (Bogen 1973: 413)

It is interesting to note that as opposed to the left hemisphere, the right hemisphere represents the whole physical body, which in itself is a universe at the microcosmic level.

Table 6.1: Dichotomies between the Left and Right Hemispheres

Left Hemisphere	Right Hemisphere
World	Self
I	Me
Doing	Being
Humanity	God
Knowledge	Knowing
Past and future	Present moment
Filtered reality	Reality as it is
Masculine	Feminine

Stress and Burnout: A Wake-up Call to the Right Hemisphere

More and more people in the civilized world suffer from stress and burnout. This tendency is escalating into epidemic proportions, with thousands and thousands of people all over the world suffering from stress and stress-related diseases, including severe emotional problems such as anxiety and depression.

Strain from work and a hectic everyday life is not the sole reason for severe stress diseases and burnout. I believe it is only a trigger. As I have said already, most people today suffer deep pain, consisting partly of repressed feelings and childhood traumas, and partly of alienation caused by a loss of connection to who they really are. They feel abandoned and incomplete, and that puts a tremendous burden on their lives. In fact the alienation seems to be the karma of the whole human race at the present level of evolution. Can we relate this malady to brain evolution and changes in hemisphere dominance?

Normally, during daily routines, the left hemisphere is in charge of behaviour and is more active than the right hemisphere. However, during unusual circumstances, the right hemisphere steps in and takes control. US neuroscientist Elkhonon Goldberg suggests that while the left hemisphere is concerned with cognitive routines, the right hemisphere is concerned with novelty. So, according to Goldberg, as soon as a novel or unusual situation arises, the right hemisphere takes over (Goldberg 2001: 43).

When the left hemisphere caves in because of stress, again the right hemisphere takes control. This is accompanied by a shift in EEG activity toward the right hemisphere. This shift may initially be experienced by the subject as depression, anxiety, chronic fatigue, attention deficits, hyperactivity, poor memory, sleep disturbances or some other symptom or disorder. These are all symptoms of a weakened left hemisphere and a suppressed, traumatized right hemisphere (and neglected body) waking up.

In my own work I have observed how stress can change the balance of EEG activity between the two hemispheres, shifting the activity to the right side. In most cases these EEG changes are reversible and are normalized when people's stress levels are reduced after treatment. This has also been the result of US professor Richard Davidson's work with stress.

This waking up of the right hemisphere is often painful, I believe, because this hemisphere has been neglected during the evolution of modern civilization. For a long time it has been repressed and exploited by the left and used as a 'store room' for all sorts of unpleasant material including repressed feelings and childhood trauma. Also, when the right hemisphere takes control of the body and nervous system following a breakdown, it is suddenly in charge of a 'bankrupt estate' and restoring it can be a very difficult process.

However, I believe that stress can also be an *evolutionary driver*. It seems that people don't change unless they have to, unless they are forced to by life circumstances. President of the Foundation for Consciousness Evolution Barbara Marx Hubbard (2007) agrees when she says that increased pressure from crises may lead to either destruction or evolution to a higher level. She even contemplates the possibility that a new species of humans is evolving.

Thus, stress and burnout can be seen as evolutionary drivers forcing the right hemisphere to take charge of behaviour.

I actually believe that stress can precipitate *kundalini* activity and eventually a *kundalini* awakening. This is also the view of Genevieve Paulson, who writes:

> *People who have overworked for years may have nervous, physical, emotional, or mental break downs and require several months' or years' hiatus to recover; many times this, too, is attributable to excessive* kundalini *pulled out by the system to handle the overload.*
>
> *(Paulson 2002: 11)*

So stress may in fact act as a facilitator in the birth of the New Brain. Increasing stress levels in people will eventually cause a shift in dominant hemisphere activity to the right side. When this happens, people will be confronted with their unconscious, and the agonizing birth of a New Brain will begin.

AN EVOLUTIONARY BRAIN MODEL: ALTERNATE HEMISPHERIC DOMINANCE

Burnout is a strong blow to the ego, but, as we have just seen, can also be a wake-up call. As bad as it looks, it may warrant a new beginning, the dawn of a new consciousness.

After a breakdown caused by stress, people usually recover, but often their personal values and sometimes even their personality change. Often they become less materialistic, more sociable, more compassionate and concerned about other people and their well-being. I believe that these personality changes are based on underlying neurological changes in the brain that could be due to a shift in brain activity from the left to the right hemisphere. Here again, I see the right brain as a 'default' mechanism, a primary brain to fall back on in times of severe stress and change.

I want to propose here a model of consciousness evolution based on an evolutionary cycle of *alternate hemispheric dominance*. My model is based on the assumption that originally, perhaps only a few thousand years ago, the right hemisphere was dominant for most functions and behaviour, as suggested by Julian Jaynes. In other words, in ancient times the right hemisphere was the prime carrier of human consciousness. Since a child's development seems to reflect the development of the human species, it should not come as a surprise that small children (up to around two years) are totally dominated by their right hemisphere. Elkhonon Goldberg (2005) says that the right hemisphere is dominant in the early years, but as the child grows the left hemisphere gradually becomes more dominant.

A gradual shift from right to left hemisphere dominance takes place in all young children at around the age of two, coinciding with the development of language functions in the left hemisphere. And I also believe that the same change of hemisphere dominance from the right to left side took place in our ancient predecessors.

According to anthropologists, many indigenous people (for example the Hopi Indians of North America and the aboriginal Australians) perceived a different reality from us. They had softer ego boundaries and felt less separate from their tribal companions and from nature. Above all, they saw nature and everything around them as alive; to them there was no such thing as an inanimate object. Steve Taylor compares the consciousness of indigenous people with an awakened consciousness when he says, 'The permanently heightened state of consciousness which the mystic strives so hard to attain was – at least to a certain level – the natural birthright of indigenous people.' (Taylor 2010: 51) However, indigenous people were not awakened. I think they belonged to an earlier stage of evolution where life was experienced almost exclusively through the right hemisphere, just as is the case in young children.

I think that in due time, the indigenous people of the world must evolve into their left hemisphere and go through the same mental turmoil which characterizes people in developed societies. They must do that in order to complete the evolutionary cycle. Then, in order to become awakened, the frontal lobes of these people must mature and evolve to a much higher level. That, at least, is my theory.

To sum up the theory in its entirety: in ancient times people were totally dominated by their right-brain hemisphere, which was then the primary carrier of their consciousness. At some point, however, certain brain functions (such as language, logic and analysis) lateralized to the left hemisphere. From then on people became more and more dependent on their left, verbal, analytical brain hemisphere. This development probably came about because life circumstances demanded advanced mental abilities such as verbal communication, abstract thinking and the ability to plan ahead of time. During the last few decades of the information age, this tendency has accelerated enormously, to the extent that most people, at least in the Western world, have got completely 'lost' in their left hemisphere, building up their egos (with which they are totally identified) and at the same time losing the connection with who they really are.

As a consequence of my theory of evolutionary, alternate, hemispheric dominance I predict that the long evolutionary voyage from the right to the left hemisphere is coming to a turning point, where a large number of people on the planet will begin to reconnect to their right hemisphere in their search for meaning and for a lost Self.

The Right Hemisphere and the Return to the Real Self

My evolutionary model suggests that the egotistic left hemisphere has overburdened the body and nervous system so much that many people have reached breaking point. You might call this the 'bankruptcy' of the ego.

Following a collapse of the ego due to stress, burnout or some other life-changing incident, many people embark on a quest to find their true, authentic Selves – Selves which were probably lost a long time back in early childhood (as suggested by Arthur Janov in his bestseller *The Primal Scream*). Since the real Self is not to be found in the external world (left hemisphere), they have to look inward (to the right hemisphere), and that, I believe, is what more and more people are doing today. I think they are trying to recover a repressed and forgotten authentic Self that still resides in their right hemisphere, latent but holding the potential of the peace, joy and excitement they haven't felt since they were small children. Whether they enter therapy, meditate or take mind-expanding substances, I think they are all searching for a lost Self.

There is a 'paradise lost' theme in the Book of Genesis which can be viewed as a parable of the initial transition from the right to the left hemisphere. Adam and Eve's eating from the Tree of Knowledge would thus symbolize humankind's first step from the right into the left hemisphere – the hemisphere of knowledge. This evolutionary capture of a new brain territory now provided humanity with conscious self-awareness and with the ability to observe itself. In the parable, Adam and Eve suddenly saw themselves as others would see them: they saw that they were naked. They had become self-conscious.

It is tempting for me to speculate that this alienation from the real Self and from God has a neurological counterpart in the shift in brain evolution from right to left-hemispheric dominance. If this hypothesis holds true, we would expect a return to right-hemisphere dominance to lead to reunion with our Self and with God.

I believe that following the collapse of the ego, it is time to return to the real Self and to recognize the primacy of the right hemisphere. Of course we still need to use the logical left brain, but it should not be running the show all by itself anymore; it should be a servant, not a master.

I submit that now is the time to open up to that huge realm of genuine peace, joy, love and compassion that I believe is latent in all of us in the right hemisphere. This is what must happen if we are to have a new and better world.

Jill Taylor agreed with that when she said:

> *My stroke of insight is that at the core of my right hemisphere consciousness is a character that is directly connected to my feeling of deep inner peace, love, joy, and compassion in the world.*
>
> *(Taylor 2008: 133)*

She went on to say:

> *I find that my right hemisphere consciousness is eager for us to take* that next giant leap for mankind and step to the right *[my emphasis] so we can evolve this planet into the peaceful and loving place we yearn for it to be.*
>
> *(Ibid: 177)*

Chapter 7
TOWARD A NEW WORLD

Over the past few decades the Earth and its inhabitants have gone through many rapid changes, which have given rise to much concern. Climate change, famine and environmental pollution constitute a real threat to humanity. In addition there is a massive exploitation of natural resources, for example cutting down the rainforest, which could threaten life on Earth. Military armament, the production of nuclear weapons, terrorism and armed conflict in many parts of the world also threaten the survival of our species. If we do not take responsibility and do something radical to resolve these problems, our children and grandchildren (if they are still around) will face much larger problems.

On the psychological and social level, humanity's greed and selfishness have led to huge economic differences between countries and between the inhabitants of any given country. The gap between rich and poor becomes wider and wider. Some people are incredibly rich, while millions, for example in Africa, starve to death or live an absolutely minimal existence.

The ego is to be held responsible for most of these calamities. The basic ego patterns, we recall, are resistance, control, power, greed, defence and attack. These are the qualities of the present state of the human mind, which will eventually lead us all to disaster unless a radical change of consciousness takes place.

In his book *A New Earth*, Eckhart Tolle explains that because of the dysfunction of the egoic, human mind, magnified through science and technology, the survival of the planet is for the first time in jeopardy. How do we meet this major challenge? Tolle writes:

> *...when survival is threatened by seemingly insurmountable problems, an individual life-form – or a species – will either die or become extinct or rise above the limitations of its condition through an evolutionary leap ... Humanity is now faced with a stark choice: Evolve or die.*
>
> *(Tolle 2005: 20–21)*

Václav Havel, former president of Czechoslovakia, had already warned the world in 1990, when addressing the US Congress:

> *Without a global revolution in the sphere of human consciousness, nothing will change for the better in the sphere of our being as humans, and the catastrophe toward which this world is headed –*

> *be it ecological, social, demographic or a general breakdown of civilization – will be unavoidable.*
>
> *(Havel 1990)*

I have no doubt that the global crisis now facing us is basically a crisis of consciousness. Human greed and selfishness have gone too far. It is time to stop the race for growth, more growth and material wealth while polluting and exploiting the Earth and suppressing millions of its inhabitants. And it is time to realize that humanity is one big family with a common destiny. What is good for our fellow human beings is also good for us and good for the whole.

Listen to Jill Bolte Taylor, the neuro-anatomist who was temporarily only able to experience the world through her right hemisphere:

> *Freed from all perception of boundaries, my right mind proclaims, 'I am part of it all. We are brothers and sisters on this planet. We are here to help make this world a more peaceful and kinder place.' My right mind sees unity among all living entities, and I am hopeful that you are ultimately aware of this character within yourself.*
>
> *(Taylor 2008: 141)*

The biggest obstacle facing the emergence of a new consciousness and a new world is the human ego, with its destructive feelings and insatiable greed. Asked what kind of change in consciousness is needed in order for humanity to survive, Stanislav Grof answered:

> *...we need both new strategies that would allow the transformation of destructive human tendencies, such as malignant aggression and insatiable greed, and a profound revision of our value system and scientific worldview.*
>
> *(Russell, Grof, Lazlo: 28)*

It makes sense to me to see the Earth as a living organism. That was actually what James Lovelock suggested in his book *Gaia*. Peter Russell elaborated on the concept. In his book *The Global Brain*, he points to the fact that the world and its inhabitants are becoming more and more interconnected. Only 100 years ago, the only means of communication was language, either spoken or written. More recently the telegraph and the telephone were invented, enabling people to communicate over long distances. Most recently, in the information age, the interconnectivity has exploded through the development of mobile phones, e-mail and the internet. Russell states that we are now in a situation 'where information can be transmitted to anyone, anywhere in the world, at the speed of light'. He goes on to say:

> *Billions of messages continually shuttling back and forth, in an ever-growing web of communication, linking the billions of minds of humanity together into a single system. Is this Gaia growing herself a nervous system? ... No longer will we perceive ourselves as isolated individuals; we will know ourselves to be a part of a rapidly integrating global network, the nerve cells of an awakening global brain.*
>
> *(Russell 2007: 107–10)*

Incidentally, as people all over the world are more and more intimately connected via the internet, Gaia becomes more and more 'present'. An injustice committed, for example, in some remote ('unconscious') area in Africa will be known to the world (for example through internet social services such as Facebook, Twitter and YouTube) almost immediately (and thus made 'conscious' to Gaia). When Gaia develops such a highly functioning 'nervous system', illegal, covert and undemocratic operations anywhere in the world will be more difficult to execute, since the general public will know about it right away. Gaia will be conscious of the event, and hopefully respond in a proper manner.

It is unfortunate that modern civilization gives its full support to many human egoistic activities, more or less accepting the motto 'Every man for himself.' I believe that this attitude is founded on the belief system that life is basically meaningless, without purpose, and that you have to try to get the most out of it for yourself, even at the cost of others. This, of course, is not true. Life is not accidental, and we must realize that everybody on this planet is united and has a common destiny.

Fortunately, I believe that many people are beginning to see the dire consequences of the ego's many selfish activities. On the financial-political level we have a world financial and economic crisis threatening our present materialistic lifestyle. The Earth is beginning to choke on its own garbage, creating global pollution, and excess energy expenditure is creating global warming and climate problems.

There is a biblical prophecy in both the Old and New Testaments, speaking of a collapse of the existing world order and the arising of 'a new heaven and a new Earth' – the Bible seems to predict that our present world order must break down before a new consciousness and a new world can arise. Does that indicate that humanity must almost destroy itself before it comes to its senses? Are we witnessing the beginning of a collapse of the civilized world order this very moment? If we look at the global financial problems, the environmental pollution, global warming, the exploitation of natural resources and of people, not to speak of the wars, armed conflicts and terrorism going on in the world today, we may certainly get the impression that Armageddon is near.

A NEW CONSCIOUSNESS AND A NEW WORLD

How would life be in a world populated by a new species of 'front-brain right-hemisphere' people? The emergence of such a New Brain will of course give rise to many changes in our experience of life and of ourselves. This is my impression of the changes that will follow the development of a new brain:

- There are strong perceptual changes: everything looks more vivid, detailed and real.
- The mind becomes very quiet. The usual incessant stream of comments and judgements about whatever we are experiencing fades away.
- There is always an inner state of peace and joy which is unshakable, no matter what happens in the outer world.

- Feelings are still there, but they are short-lived and there is no identification with either feelings or thoughts.
- Our old personality is still there, but we do not identify with it anymore and can observe it from a higher level.
- Old belief systems fade away.
- Love, compassion and a willingness to help others are predominant and it becomes difficult, if not impossible, to hurt other people and animals.
- All interest in the past and the future is lost, since all focus is on the present moment.
- Most actions and behaviour are not preceded by thinking but guided by intuition.
- There is a strong feeling of interconnectedness and empathy between people, and between mankind and Mother Earth. Eventually people will develop a feeling of oneness with the Earth and with everything on it. They will feel they are part of a huge living organism.
- There will be a shift of values away from materialism and competitiveness, toward more playfulness, helpfulness, creativity and empathy.

To come back to the world as it is now, in my view natural disasters (earthquakes, flooding, etc.) killing thousands of people may only reflect severe conflicts at the present level of human consciousness. What I am suggesting is that there is a connection between the level of human consciousness and the condition of the physical world we live in. Indian master Bhagavan Kalki says:

> *Man will soon realize that the earth is a living organism, he depends on like his mother... There is a very close correlation between human consciousness and the physical processes occurring on the planet. So the moment that conflict levels are reduced in human consciousness, you will find dramatic changes at the earth level also.*
>
> *(Quoted Windrider, www.spiritwheels.com)*

This view is in accordance with the concept of Gaia, the Earth as a living organism.

Our present society is to a large extent fear-driven. I think that in the new world there will be little fear. Jill Taylor gives an illustration of how she experienced the world from her right-hemisphere perspective:

> *How on earth would I exist as a member of the human race with this heightened perception that we are each a part of it all, and that the life-force energy within each of us contains the power of the universe? How could I fit in with our society when I walk the earth with no fear?*
>
> *(Taylor 2008: p.70).*

Is it possible today to imagine a world without war, terrorism, crime, sickness and starvation, a world without fear, greed, and exploitation, a peaceful world with redundant resources where most people have enough and very few have too little? Is such a new world possible? I think it is, but we must have the vision and the will to see it and believe that it is possible.

A NEW SPIRITUAL SCIENCE

I suggest that in the new world there will be a gradual transition from religion to a spirituality based on logic, reason and science. Today, many people are turning away from religion, as they see it as obsolete or superstitious. Most religions are easy subjects for criticism, since in their present form they often present a diluted and distorted version of an original message of the eternal wisdom. Also, the dire consequences of religious fundamentalism have put religion in a bad light.

It seems appropriate here to distinguish between organized religion and spirituality. There is a big difference. Religions ask you to adopt their values and belief systems and to learn from others' experience, while spirituality urges you to seek your own values and truth. Religions most often reject spirituality, since it can lead to a different conclusion from that of the religion itself.

Many people turn away from religion because they can no longer, in blind faith, accept more or less outdated teachings, and would like to have, if possible, a logical, reasonable explanation of the relationship between human consciousness, God and the universe. I see this book, among many others, as an attempt to promote a new spiritual science for modern, educated people.

We hear from our most prominent scientists that God is a delusion, the universe created itself and life occurred by chance. Since science supposedly represents the highest knowledge and authority in modern society, its negative materialistic attitude toward spirituality constitutes a serious problem for people who are considering following a spiritual path. Stanislav Grof says:

> *In a culture where science enjoys great respect and authority, if its message is distinctly anti-spiritual, it seriously inhibits people's interest in the spiritual quest.*
>
> *(Russell, Grof, Lazlo: 29)*

How can spiritual science deal with the puzzle of God and the problem of creation? Recently an interesting book was published: *The God Theory* by astrophysicist Bernard Haisch. Haisch argues that there is an eternal all-encompassing omnipresent Consciousness that is God, to whom we are all connected via our individual consciousness. God's Consciousness is based on all living beings in the universe, thus we are all (plants, animals and humans) carriers of it and our experiences are also God's experiences. 'The point of a created universe is to experience it. Life is God made manifest,' says Haisch (2006: 22), who also believes that Consciousness is primary to the physical world: 'It is my contention, and the crux of the God Theory, that ideas created by a spiritual Consciousness are the cause and basis of the physical world.' (Ibid: 58) This is another way of saying that the universe is the body of God. According to Haisch, there is nothing in modern science that contradicts the 'God Theory'. The only difference between the God Theory and modern science is that Consciousness pre-exists.

The basic problem modern science has in understanding creation is the deep-rooted belief that something can evolve out of nothing. Haisch: 'Creation of something out of nothing is impossible and God is all there is.' (Ibid.) But is it possible to say anything about how the universe was created – how it all started? According to Haisch, originally the unmanifest God existed as an all-pervading Consciousness. But in order to know (experience) himself He had to split himself up into a part that saw and another part that was seen. At this point, the One became the many and they were all manifestations of God's Consciousness.

At some point in evolution, in order to experience life to its fullest humans had to leave paradise and be alienated from their true nature and from God. This was all part of a divine plan enabling humans (and God) to experience life to its utmost potential. Listen to what Ken Wilber, a contemporary spiritual thinker, said in a dialogue with Andrew Cohen, another spiritual teacher and thinker:

> *...you can't go through the whole process of evolution knowing you are God. That's just not going to work. So you would have to forget who you are; you'd have to get lost – convincingly lost – or it's not a game and it's no fun at all! So you get lost and then you slowly awaken... And then at some point evolution is going to become self-conscious, and then it's going to become super-conscious. But it's taken fourteen billion years to get to this point.*
>
> *(www.enlightennext.org/magazine/j40/guru-pandit.asp.)*

SUMMARY AND CONCLUSIONS

From what we have seen so far in this book, it appears that the brain itself does not cause Consciousness, it is only a vehicle for the experiencing Consciousness. The common notion that Consciousness emerges from physical matter during evolution as a result of sufficient brain mass and neuronal complexity is not logical. I feel sure that an eternal, omnipresent, experiencing Consciousness (some call it God) has always existed – it pre-exists everything else.

It is also mandatory to distinguish Consciousness from the human mind. While the human mind is undergoing an evolution, the experiencing Consciousness is not – it is eternal and unchangeable.

The human brain is still undergoing an evolution toward higher levels of functioning. I foresee that a New Brain will develop in humans. This new level of brain organization will include more energized and active frontal lobes, together with an opening up of more neural connections between the cerebral hemispheres. This will lead to increased activity in the right hemisphere, which eventually will become better integrated with the left, logical hemisphere and an equal partner with it.

In order to attain higher consciousness, the frontal lobes must be charged with more energy. According to yogic philosophy, this energy must come from *kundalini*, a bioelectric energy residing at the base of the spine. When this energy is aroused, either spontaneously or through other means, it ascends through the spinal column, and when it reaches the brain it is called a *kundalini* awakening.

There is no direct measure of *kundalini* energy. However, I have shown that an indirect measure, such as frontal gamma wave activity, may be used for identifying *kundalini* activity. Many people who have been diagnosed with a psychiatric disorder may instead be suffering from a complicated *kundalini* awakening. Knowledge of *kundalini* and proper guidance could work wonders with these people. Therefore I think it is crucial that science makes a serious effort to study the *kundalini* phenomenon.

In my studies of altered states of consciousness (induced by meditation, feeling release therapy and the intake of ayahuasca) using electroencephalography (EEG) and brain mapping, I have found that altered states provide access to the unconscious and primarily involve the right cerebral hemisphere but not the frontal lobes.

On the contrary, awakened or higher states of consciousness primarily involve the frontal lobes and only secondarily the right cerebral hemisphere.

Awareness is a driving force in consciousness development. Meditation is the best-known and most widespread form of awareness training. However, a new technology called neurofeedback or brainwave training can be very effective in improving body awareness, focus and presence. It may also be used for stimulating creativity, flow and intuition and can be a preparation for higher states of consciousness.

Access to the right hemisphere and a 'clean-up' of the unconscious can take place through meditation, alpha wave training, feeling release therapy or drinking ayahuasca. These methods do not automatically lead to a higher state of consciousness but may be regarded as preparation.

The frontal lobes of the New Brain can be activated through a *kundalini* awakening but can also be stimulated via certain types of meditation and frontal gamma wave training.

It is important to distinguish between an altered state of consciousness and a higher state of consciousness. These states are based on different brain structures and give rise to different experiences. In altered states you are confronted with the unconscious and all its weird phenomena, while in an awakened state you see reality as it is: very lucid and very real.

Going into the highest state of consciousness, called *samadhi*, will bring about a return to the Creator. Thus *samadhi* is creation in reverse. In it there is no subject to experience, since the subject is included in the experience. *Samadhi* can only be realized in retrospect, when the subject returns to normal consciousness. It is the experience of pure Consciousness, Nothingness, the Void or God.

I have suggested a model of *alternate hemispheric dominance*, according to which consciousness moves from the right to the left brain during the process of evolution and eventually back to the right hemisphere. I believe that the long evolutionary voyage from the right to the left brain, which started thousands of years ago, is now coming to a turning point. Following the surrender of the ego (left hemisphere), it is time to reconnect with the right hemisphere and return home to our true Self.

I have no doubt that the global crisis now facing us in the form of huge financial, ecological and social problems is basically a crisis of consciousness. It is the human ego with its destructive thoughts, feelings and insatiable greed that stands in the way of a new and better world.

The evolution of a New Brain and the transformation of human consciousness to include more love, empathy and compassion for fellow human beings are necessary in order to create a new and better world and to avoid the destruction of civilization.

In the new world I believe organized religion will be substituted by a spirituality based upon logic, reason and science. I think this will happen because modern, educated people cannot believe in ancient religious dogma without a logical explanation. Such a spiritual science could help people to realize that life as we know it is not a meaningless, mechanical accident in a remote corner of the universe. On the contrary, life has both meaning and purpose and is guided by a higher intelligence.

It is my hope that before the dire consequences of humanity's egoistic and harmful behaviour annihilate civilization, humans will take the next step in evolution and mobilize enough will and presence to take action in order to bring Mother Earth back on the right track toward a new and better world.

BIBLIOGRAPHY AND RECOMMENDED READING

A Kosmic Roller-Coaster Ride: Andrew Cohen and Ken Wilber in dialogue. www.enlightennext.org/magazine/j40/guru-pandit.asp.

Bagchi BK, Wenger MA (1957). Electro-physiological correlates of some Yogi exercises. *EEG and Clinical Neurophysiology* 9. 51.

Beauregard M, O'Leary D (2007). *The Spiritual Brain: A neuroscientist's case for the existence of the soul*. New York, NY: HarperCollins.

Bentov I (1977). *Stalking the Wild Pendulum*. Rochester, VT: Bear & Company. www.innertraditions.com

Bogen JE (1973). *Proceedings of the Society of Neuroscience* 3. 413.

–, Bogen GM (1969). The other side of the brain. 3. The corpus callosum and creativity. *Bulletin of the Los Angeles Neurological Society* 34(4): 191–220. www.its.caltech.edu/~jbogen/text/OSOB_3.html

Bucke RM (1901). *Cosmic Consciousness: A study in the evolution of the human mind.* Philadelphia, PA: Innes & Sons.

Busch NA, Dubois, J, VanRullen R (2009). The phase of ongoing EEG oscillations predicts visual perception. *Journal of Neuroscience* 29(24): 7,869–76.

Chalmers D (1996). *The Conscious Mind*. Oxford: Oxford University Press.

Chi R, Snyder A (2011). Electric thinking cap? Flash of fresh insight by electrical brain stimulation. www.sciencedaily.com. 2 February.

Cohen HC, Rosen RC, Goldstein L (1976). Electroencephalographic laterality changes during human sexual orgasm. *Archives of Sexual Behavior* 5(3):189–99.

Crews DJ, Landers DM (1993). Electroencephalographic measures of attentional patterns prior to the golf putt. *Med Sci Sports Exercise* 25(1): 116–26.

Crick F (1995). *The Astonishing Hypothesis: The scientific search for the soul*. New York, NY: Simon & Schuster.

Csikszentmihalyi M (1991). *Flow: The psychology of optimal experience*. New York, NY: Harper & Row.

Dalai Lama (2005). *The Universe in a Single Atom*. New York, NY: Morgan Road Books

Das NN and Gastaut H (1955). Variations de l'activité électrique du cerveau dans la meditation et l'extase yoguique. *EEG and Clinical Neurophysiology* 6: 211–19.

Dass, R (2004). *Paths to God: Living the Bhagavad Gita*. New York, NY: Crown Publishing/Random House.

Davenport TH, Beck JC (2001). *The Attention Economy: Understanding the new currency of business*. Cambridge, MA: Harvard Business School Press.

Edelstein A (2000). When you go beyond the ego you become an offering to the world: interview with Amma. *What is Enlightenment?* 17. Summer. 34. www.enlightennext.org/magazine/j17/amma.asp.

Egner T, Gruzelier, JH (2003). Ecological validity of neurofeedback: modulation of slow wave EEG enhances musical performance. *NeuroReport* 14(9): 1,221–4.

Fischer R, Rhead, J (1974). Nature, nurture and cerebral laterality. *Confinia Pschiatrica* 17(3–4): 192–202.
Basel: S. Karger AG.

Foulkes D (1966). *The Psychology of Sleep*. New York, NY: Charles Scribner's Sons.

Fröhlich H, Kremer F (1983). *Coherent Excitations in Biological Systems*. Heidelberg: Springer-Verlag.

Goldberg E (2001). *The Executive Brain: The frontal lobes and the civilized mind.* Oxford: Oxford University Press.

– (2005). *The Wisdom Paradox*. New York, NY:Free Press.

Goldstein L, Sugerman (1969). EEG correlates of psychopathology. Zubin and Shagass (eds). *Neurobiological Aspects of psycho-pathology*.

Goldstein L, Stoltzfus N (1973). Psychoactive drug-induced changes of interhemispheric EEG amplitude relationships. *Agents and Actions*. April.

Goleman D (2003). *Destructive Emotions: A scientific dialogue with the Dalai Lama*. New York, NY: Bantam Dell.

Goswami A (2001). *Physics of the Soul*. Charlottesville, VA: Hampton Roads.

Grob C (1999). The psychology of ayahuasca. *Ayahuasca: Hallucinogens, consciousness and the spirit of nature*. R Metzer (ed.) New York, NY: Thunder's Mouth Press.

Haisch B (2006). *The God Theory*. San Francisco, CA: Red Wheel/Weiser.

– (2010). *The Purpose Guided Universe*. Pompton Plains, NJ: New Page Books.

Harrington A, Zajonc A, eds (2006). *The Dalai Lama at MIT*. Cambridge, MA: Harvard University Press.

Havel V (1990). http://everything2.com/node/851021

Hoffmann, E (1980). EEG Frequency Analysis of Anxious Patients and of Normals during Resting and during Deep Muscle Relaxation. *Adv. Biol. Psychiat.* 4: 124.

– (1998). Mapping the brain's activity after Kriya Yoga. *BINDU* 12. http://www.yogameditation.com/Articles/Issues-of-Bindu/Bindu-12/Mapping-the-brains-activity-after-Kriya-Yoga.

Hoppe, KD (1989). Psychoanalysis, hemispheric specialization, and creativity. *Journal of the American Academy of Psychoanalysis*. 17(2): 253–69.

Huxley A (1954). *The Doors of Perception*. New York, NY: Harper & Row.

Irving D (1995). *Serpent of Fire: A modern view of kundalini*. York Beach, ME: Samuel Weiser.

James W (1890, 2007). *The Principles of Psychology*. Vol.1. New York, NY: Cosimo, Inc.

Janov A (1970). *The Primal Scream*. New York, NY: Dell Publishing.

– (1991). *The New Primal Scream*. Wilmington, DE: Enterprise Publishing.

– (1996). *Why You Get Sick, How You Get Well*. Garden City, MI: Dove Books.

Jaynes J (1976). *The Origin of Consciousness in the Breakdown of the Bicameral Mind*. Boston, MA: Houghton Mifflin.

Jeffreys J. New research clarifies human understanding provides a solution to the elusive binding' problem. www.web-us.com/40hz/WW124.htm.

Joseph R (1992). *The Right Brain and the Unconscious: Discovering the stranger within.* New York, NY: Plenum Press.

Kalki B. www.onenessuniversity.org.

Kieffer G (1983). Interview with Gopi Krishna. *Yoga Journal*. November–December.

Krishna G (1975). *The Awakening of Kundalini*. Markdale, ON: Institute for Consciousness Research.

– (1993). *Living with Kundalini: The autobiography of Gopi Krishna*. Boston, MA: Shambhala Publications.

Laszlo E, Grof S, Russell P (2003). *The Consciousness Revolution*. Las Vegas, NV: Elf Rock Productions.

Lovelock JE (1979). *Gaia: A new look at life on Earth*. Oxford: Oxford University Press.

Lutz A *et al.* (2004). Long-term meditators self-induce high-amplitude gamma synchrony during mental practice. *Proceedings of the National Academy of Science*. November.

MacLean PD (1973, 1990). *The Triune Brain in Evolution*. New York: Springer.

Martinus (1939). *Book of Life* (Danish: *Livets Bog*). Copenhagen: Martinus Institute.

Marx Hubbard B *et al.* (2007). A Vision for Humanity. *The Mystery of 2012*. Boulder, CO: Sounds True.

Metzner R, ed. (1999). *Ayahuasca: Hallucinogens, consciousness and the spirit of nature*. New York, NY: Thunder's Mouth Press.

Muktananda S (1979). *Kundalini: The secret of life*. New York, NY: Syda Foundation.

Murali Krishna R. *The Mind of a Mystic*. Oklahoma Research report. Oklahoma City, OK: lifeblissfoundation.org/founder_science_spirituality.asp.

Nagata K (1988). Topographic EEG in brain ischemia: correlation with blood flow and metabolism. *Brain Topography* 1. 97–106.

Newberg AB (2004). This is your brain praying. *Spirituality and Health*. January.

–, D'Aquili E (2001). *Why God Won't Go Away*. New York, NY: Ballantine Books.

–, Iversen J (2003). The neural basis of the complex mental task of meditation. *Medical Hypotheses* 61(2): 284.

Osho (1976). *The Art of Dying: Talks on Hasidism*.

– (1976). *Meditation: The art of ecstasy*. Osho International Foundation

– (1984). *In Search of the Miraculous*. Rajneeshpuram, OR: Rajneesh Foundation International.

– (2001). *Intuition: Knowing beyond logic*. New York, NY: St. Martin's Griffin.

– (2001). The Reality of Dreams. *Osho Times*. January.

– (2005). *Yoga: The path to liberation*. India: Penguin Books.

Paulson GL (2002). *Kundalini and the Chakras*. Woodbury, MN: Llewellyn Publications.

Pink DH (2005). *A Whole New Mind: Why right-brainers will rule the future*. New York, NY: Riverhead Books.

Persinger M (2003). Experimental Simulation of the God Experience. *NeuroTheology: Brain, science, spirituality, religious experience*. R. Joseph. San José, CA: University Press of California.

Rizzolatti G, Craighero L (2004). The mirror-neuron system. *Annu Rev. Neurosci. 27: 169–92.*

Russell P (2005). Mathematics and reality. http://www.peterrussell.com/Reality/realityart.php#Mathematics. 16 January.

– (1992). *Waking Up in Time*. San Rafael, CA: Origin Press.

– (2002). *From Science to God: A physicist's journey into the mystery of consciousness*. Novato, CA: New World Library.

– (2003). Deep mind: beyond science, behind spirit. *Resurgence.* October. www.peterrussell.com.

– (2007). *The Global Brain: The awakening Earth in a new century*. Edinburgh: Floris Books.

Sanella L (1992). *The Kundalini Experience: Psychosis or transcendence?* Lower Lake, CA: Integral Publishing.

Satyananda S (1996). *Kundalini Tantra*. Bihar, India: Bihar School of Yoga. Second edition.

Schwartz JM, Begley S (2003). *The Mind and the Brain: Neuroplasticity and the power of mental force*. New York, NY: HarperCollins.

Sheth BR, Sandkühler S, Bhattacharya J. (2009). Posterior beta and anterior gamma oscillations predict cognitive insight. *J Cogn Neurosci* 21(7): 1,269–79.

Simon HA (1971). Designing organizations for an information-rich world. *Computers, Communication, and the Public Interest*. M Greenberger (ed.) Baltimore, MD: The Johns Hopkins Press.

Smith AL, Tart CT (1998). Cosmic conscious experience and psychedelic experience. *Journal of Consciousness Studies* 5(1): 100–4.

Sperry RW (1974). Lateral specialization in the surgically separated hemispheres. In FO Schmitt, FG Worden (eds). *The Neurosciences* 3rd Study Program. Cambridge, MA: MIT Press.

Stolaroff M (1979). LSD – Twenty Years After. www.hofmann.org/papers/LSD20YEARSLATER.htm.

Tart CT (1987, 2001). *Waking Up*. Lincoln, NE: iUniverse.com.

Taylor JB (2008). *My Stroke of Insight: A brain scientist's personal journey*. London: Hodder & Stoughton.

Taylor S (2010). *Waking from Sleep*. London: Hay House.

– (2011). *Out of the Darkness*. London: Hay House.

Tolle E (1999). *The Power of Now*. Novato, CA: New World Library.

– (2005). *A New Earth*. London: Penguin Books.

Wheeler R, Davidson R, Tomarken A (1993): Frontal brain asymmetry and emotional reactivity. *Psychophysiology* 30. 82–9.

Windrider K. www.onenessForAll.com.

– Conversation with Bhagavan. www.spiritwheels.com.

ABOUT THE AUTHOR

Erik Hoffmann has a master's degree in psychology from Copenhagen University in Denmark where he worked as an assistant professor at the Psychological Institute for eight years. His research was in managing stress with biofeedback, and studied altered states of consciousness using electroencephalography (EEG). In several periods he served as a visiting professor at Rutgers University, New Jersey, where researched computerized EEG studies of brain functions in psychiatric patients. His research results have been published in scientific journals and presented at international brain conferences.

From 1994–6 Erik worked at the International Primal Center in Los Angeles, California, where he did an EEG brainmapping study of the effects of feeling release therapy. The results of this research were in February 1995 presented to the Californian Psychological Society in La Jolla, California.

In recent years Erik has studied meditation and altered states of consciousness using EEG methods and brainmapping. This work took place in Sweden, the Netherlands, Brazil and India, and the results were presented at international brain conferences. In the summer of 2001 he founded with Inger Spindler Mental Fitness & Research Centre in Symbion Science Park, Copenhagen. He worked there as a research director, brainwave training and brainmapping children with ADHD and adults with stress problems. Erik has been semi-retired since 2007.

www.newbrainnewworld.com

We hope you enjoyed this Hay House book.
If you would like to receive a free catalogue featuring additional Hay House books and products, or if you would like information about the Hay Foundation, please contact:

Hay House UK Ltd
292B Kensal Road • London W10 5BE
Tel: (44) 20 8962 1230; Fax: (44) 20 8962 1239
www.hayhouse.co.uk

***

Published and distributed in the United States of America by:
Hay House, Inc. • PO Box 5100 • Carlsbad, CA 92018-5100
Tel: (1) 760 431 7695 or (1) 800 654 5126;
Fax: (1) 760 431 6948 or (1) 800 650 5115
www.hayhouse.com

***

Published and distributed in Australia by:
Hay House Australia Ltd • 18/36 Ralph Street • Alexandria, NSW 2015
Tel: (61) 2 9669 4299, Fax: (61) 2 9669 4144
www.hayhouse.com.au

***

Published and distributed in the Republic of South Africa by:
Hay House SA (Pty) Ltd • PO Box 990 • Witkoppen 2068
Tel/Fax: (27) 11 467 8904
www.hayhouse.co.za

***

Published and distributed in India by:
Hay House Publishers India • Muskaan Complex • Plot No.3
B-2 • Vasant Kunj • New Delhi - 110 070
Tel: (91) 11 41761620; Fax: (91) 11 41761630
www.hayhouse.co.in

***

Distributed in Canada by:
Raincoast • 9050 Shaughnessy St • Vancouver, BC V6P 6E5
Tel: (1) 604 323 7100
Fax: (1) 604 323 2600

***

Sign up via the Hay House UK website to receive the Hay House online newsletter and stay informed about what's going on with your favourite authors. You'll receive bimonthly announcements about discounts and offers, special events, product highlights, free excerpts, giveaways, and more!
www.hayhouse.co.uk

CPSIA information can be obtained at www.ICGtesting.com
Printed in the USA
LVOW041335140312

273047LV00001B/2/P